Measuring Time

International Series in Operations Research & Management Science

Volume 136

For other titles published in this series, go to
www.springer.com/series/6161

Mario Vanhoucke

Measuring Time

Improving Project Performance Using Earned Value Management

Springer

Mario Vanhoucke
Fac. Economics & Business Administration
Ghent University
Tweekerkenstraat 2
9000 Gent
Belgium

Vlerick Leuven Gent Management School
Reep 1
9000 Gent
Belgium
mario.vanhoucke@ugent.be

ISBN 978-1-4614-2537-3 e-ISBN 978-1-4419-1014-1
DOI 10.1007/978-1-4419-1014-1
Springer Dordrecht Heidelberg London New York

Softcover reprint of the hardcover 1st edition 2009

Printed on acid-free paper

Springer is part of Springer Science+Business Media (www.springer.com)

The only reason for time is so that everything doesn't happen at once.

Albert Einstein

Preface

Project scheduling began as a research track within the mathematical field of Operations Research in order to mathematically determine start and finish times of project activities subject to precedence and resource constraints while optimizing a certain project objective (such as lead-time minimization, cash-flow optimization, etc.). The initial research done in the late 1950s mainly focused on network based techniques such as CPM (Critical Path Method) and PERT (Programme Evaluation and Review Technique) which are still widely recognized as important project management tools and techniques.

From this moment on, a substantial amount of research has been carried out covering various areas of project scheduling (e.g. time scheduling, resource scheduling, cost scheduling). Today the project scheduling research continues to grow in the variety of its theoretical models, in its magnitude and in its application. While the research has expanded over the last decennia, leading to project scheduling models with deterministic and stochastic characteristics, single- and multi-mode execution activities, single and multiple objectives, and a wide variety of resource assumptions, the practitioners and software tools mainly stick with the often basic project scheduling principles. This can probably be explained by the limited capability of a project schedule to cope with the uncertainty that characterizes the real life execution of the project. Indeed, the benefits of a resource-constrained project schedule have been questioned by many practitioners, and the effort someone puts into the development of a project schedule is often not in line with the benefits. Moreover, "a project schedule will change anyway due to circumstances" is often a widely used excuse to skip this important step in the project life cycle.

Nevertheless, project scheduling and project control have always been topics of interest to me ever since the research performed in my PhD period. In order to appreciate the importance of a project schedule, it should be generally accepted that the usability of a project schedule is rather limited and only acts as a point of reference in the project life cycle. Consequently, a project schedule should especially be considered as nothing more than a predictive model that can be used for resource efficiency calculations, time and cost risk analysis, project tracking and performance measurement, and so on. Throughout the years of study, both in an academic setting and in a more consultancy oriented environment, I discovered that the use of a baseline schedule is of crucial importance for project tracking, project performance measurement and schedule risk analysis. This idea silently brought me to earned value management (EVM) and arose my attention to the recent research done on this topic. The contacts and joint research interest I shared with Stephan Vandevoorde since many years, the meetings with Walt Lipke and Kym Henderson in London and the start-up of our company OR-AS together with Tom Van Acker brought everything in an acceleration. Since then, I continued doing research on fictitious and practical projects using earned value management for which the main results are written and summarized throughout the various chapters of this book.

Scope

In writing this book, I had no intention whatsoever to compete with the current excellent books of references about earned value management. Instead, the aim of this book is to throw a critical eye on the existing and newly developed techniques on EVM that measure and forecast the duration of a project. More precisely, the scope of this book can be summarized as follows:

- An overview: The book brings an overview of the common and often confusing terminology of earned value management. In this respect, many parts of this book are no more than a careful collection of statements, conclusions and results on project duration forecasting summarized from the academic and popular press.
- Formulas: The book focuses on the often simple calculations behind EVM systems rather than on the implementation details, the advantages and disadvantages and the possible impediments of these systems in practice. During the many consultancy projects, I discovered that, maybe due to the simplicity of many EVM calculations, the EVM metrics are often misunderstood or used and interpreted in a wrong way. In presenting many example calculations on small fictitious projects, I aim to bring clarity on this issue by allowing the reader to calculate along with me.
- Based on academic research: Many parts of this book are the results of academic research at Ghent University (Belgium) and Vlerick Leuven Gent Management School (Belgium). Hence, it offers a critical view on existing as well as novel EVM approaches by testing many alternative methods on a very diverse set of artificial project data that is used throughout many other, non-EVM research applications. The reader will often be referred to the current state-of-the-art literature and I truly hope that these references make the less popular academic literature a little bit more accessible to the broad audience.
- Inspired by practice: Most, if not all, results of this book are based on practical illustrations in companies, numerous discussions with colleagues and friends in charge of managing projects and by an overwhelming amount of (often virtual) discussions with project management practitioners.
- Limitations: The scope is restricted to a study on duration forecasting of a project, and hence, excludes the overwhelming amount of literature and work done on cost forecasting. The latter has been extensively investigated by, among many

others, David S. Christenson (for more information, visit the earned value bibliography[1]).

- Novel non-proven concepts: This book clearly focuses on recent research trends in earned value based duration forecasting and often brings newly developed concepts that are only recently discussed in the popular research press. It is not the intention to favor or reject any of these novel methods, but rather to (try to) bring an objective opinion by testing alternative approaches on the same project data. In this respect, the book can be used as a guideline for practitioners, and can be considered as a modest attempt to objectively compare alternative or competing EVM forecasting metrics, while keeping in mind that the ultimate truth will not be given by the formulas and simulations presented in this book.

Acknowledgements and authors

I am indebted to many people who have helped me in writing this book. First, I want to express my gratitude to Tom Van Acker (OR-AS) and Stephan Vandevoorde (Fabricom Airport Systems). Back to 2003, Stephan launched the idea to critically review the existing EVM methods in order to be able to see the bunch by the trees. Since then, he kept the research going throughout the years by guiding the many fruitful e-mail discussions between various EVM practitioners in Europe, US and Australia. Together with Tom, we have programmed our project scheduler ProTrack which is presented in chapter 7 of this book. After two years of weekend discussions and nights of programming troubles, we are proud on both our excellent cooperation and the product ProTrack that is the result of it. I am also much indebted to Walt Lipke and Kym Henderson for the many virtual and real meetings we had during the past several years, and to Ray Stratton for his quick and valuable comments on parts of this book. A special thanks goes to Broos Maenhout who has carefully read and recalculated all mathematical details of the chapters. Last but certainly not least, my sincere thanks goes to my family, especially Gaëtane for carefully reading and editing all chapters of this book, and Joyce and Thierry for their patience and their never-ending support.

The research discussed in the chapters of this book are obviously based on the common knowledge discussed throughout the literature. I want to express my gratitude to many authors that have written something in the field of project tracking in general and earned value in particular. In the remaining of this preface, I want to particularly mention a number of sources (both books and internet sites) that were helpful to me during the research project of this book. Obviously, this list does not contain an exhaustive summary of interesting references, but rather serves as a limited illustrative collection of sources useful to me and hopefully to the reader of this book.

References

Excellent books on earned value management have been reported in the literature. The books mentioned below belong to my favorites and deserve a note of

---

[1] www.suu.edu/faculty/christensend/ev-bib.html

attention since they are not all explicitly mentioned throughout the remainder of this book.

- Earned Value Project Management, 3rd Edition by Quentin W. Fleming and Joel M. Koppelman
- Practice Standard For Earned Value Management by the Project Management Institute
- Using Earned Value: A Project Manager's Guide by Alan Web
- Earned Value Management Using Microsoft Office Project: A Guide for Managing Any Size Project Effectively by Sham Dayal
- The Earned Value Management Maturity Model by Ray W. Stratton
- Earned Value Management by Roland Wanner
- Performance-Based Earned Value (Practitioners) by Paul Solomon and Ralph Young
- A Practical Guide to Earned Value Project Management by Charles I. Budd
- EVM Demystified: An Easy Guide for the Practical Use of Earned Value Management by Esther Burgess and Ruth Mullany
- Integrated Cost and Schedule Control in Project Management by Ursula Kuehn

Interesting sites
I particularly want to mention three interesting sites:

- www.earnedschedule.com: This site has been developed by Walt Lipke and is *the* site where you can find the latest developments and news about the progress in earned schedule. The site brings you the recent presentations and publications in the Measurable News and other journals and provides links to interesting contacts. With more than 13,000 hits per month in 2007, only one year after its introduction, the site can be considered as an enormous success.
- www.or-as.be: This is the site of our company OR-AS and is relevant for the reader for two main reasons. First, the reader can freely download all data files used in the simulation studies of chapters 4, 5 and 6. Moreover, the site also directs you to the software tool ProTrack which is the first and, to the best of our knowledge, only software tool which incorporates earned schedule in a traditional scheduling environment. Have fun!
- www.pmi-belgium.org: Being a Belgian citizen and having a professional career of more than 10 years in project management and scheduling naturally brings me to the Belgian chapter of the Project Management Institute (PMI) website (www.pmi.org). I want to use this opportunity to mention and promote the Belgian chapter of PMI, since many of the voluntary people have stimulated me in my research and in writing this book. Not only the financial support, but also the flow acceleration in the earned schedule interest after the chapter meeting of June 12th, 2007, have motivated me to continue the research and to write this summary book.

Awards
On June 12th, 2007, the research topic described in this book was awarded the Research Collaboration Fund by PMI Belgium. The introduction of this award was to

promote Belgian project management research, to translate results into white papers, available to all PMI members and to promote PMI Belgium outside the borders.

On November 11th, 2008, the research results have been awarded on the 22nd IPMA World Congress held in Rome (Italy). The IPMA Research Awards aim to promote excellent research to enhance project management. With these annual awards, IPMA recognizes recent outstanding contributions to the development of the discipline and profession project management through professionally conducted research. The award nomination announcement is posted on the IPMA website (www.ipma.ch) and a more detailed research description is available on www.or-as.be.

Ghent, January 2009 *Mario Vanhoucke*

Introduction

Earned Value Management systems have been setup to deal with the complex task of controlling and adjusting the baseline project schedule during execution, taking into account project scope, timed delivery and total project budget. It is a well-known and generally accepted management system that integrates cost, schedule and technical performance and allows the calculation of cost and schedule variances and performance indices and forecasts of project cost and schedule duration. The earned value method provides early indications of project performance to highlight the need for eventual corrective actions.

Although numerous excellent books and papers have been written to summarize various aspects of Earned Value Management, I believe that this book is unique in its kind and highlights earned value in a way that is different from the approach taken in any other traditional EVM book. This book is not an introductory book to EVM, nor a tutorial book on how to implement an EVM system in a company. Instead, it can be considered as a supplement on top of many other excellent books and hence, a basic knowledge about earned value will be considered as a given. I believe that this book differs in two aspects on the traditional EVM books, as follows:

1. Although earned value management systems have been proven to provide reliable estimates for the follow-up of cost performance within certain project assumptions, they often fail to predict the total duration of the project. Earned value management was originally developed for cost management and has not widely been used for forecasting a project's duration. However, recent research trends show an increase of interest to use performance indicators for predicting the total project duration. This book is a summary of a large research study that aims at validating EVM methods to forecast the total duration of a project.
2. Earned value has always been the domain of the practitioner who is in charge of managing and controlling projects. Hence, little or no effort has been done to critically analyze the behavior of EVM calculations for a wide set of very diverse project networks. This book takes a more academic approach and tests the behavior of EVM metrics on a large set of artificial data rather than on a

small sample of real data. The results on real data used in chapter 3 serve as an illustration and not as a proof of validity or general conclusion.

The book consists of eight chapters which can be briefly summarized along the following lines.

Chapter 1 gives an overview of the common and often confusing terminology of earned value management. The purpose of this chapter is twofold. First, it compares the classic earned value performance indicators SV (Schedule Variance) and SPI (Schedule Performance Index) with the newly developed earned schedule performance indicators SV(t) and SPI(t). Next, it presents a generic schedule forecasting formula applicable in different project situations and compares three methods from literature to forecast the total project duration.

Chapter 2 critically reviews and tests a novel EVM extension, the so-called p-factor approach, to measure schedule adherence based on the traditional earned value metrics. The purpose of this chapter is twofold. First, the chapter discusses the relevance of the p-factor for the detection of project impediments and/or portions of work performed under risk, based on the calculation of the traditional earned value metrics. Second, the chapter critically discusses the contribution of the p-factor to modify and improve the accuracy of the forecasts along the life of the project. Simulation results will be presented in the simulation study of chapter 4.

Chapter 3 presents a case study for three real life projects at Fabricom Airport Systems. This chapter serves as an illustration for the various concepts introduced in the previous chapters. To the best of my knowledge, this is the first time the earned schedule concept, discussed in chapter 1, is used in a practical setting in Belgium.

Chapter 4 extensively reviews and evaluates earned value based methods to forecast the total project duration based on a large Monte-Carlo simulation study. The simulation carefully controls the level of uncertainty in the project, the influence of the project network structure on the accuracy of the forecasts, and the time horizon where the earned value based measures provide accurate and reliable results. It assumes a project setting where project activities and precedence relations are known in advance and does not consider fundamentally unforeseeable events and/or unknown interactions among various actions that might cause entirely unexpected effects in different project parts. This is the first study that investigates the potential of a recently developed method, the earned schedule method, which improves the connection between earned value metrics and the project duration forecasts.

Chapter 5 sheds light on another time dimension of project management. The chapter reviews the basic calculations to measure the sensitivity of an individual activity of the project network. The relation between forecast accuracy and project sensitivity is discussed in detail. This chapter investigates the ability of activity sensitivity information to improve the project tracking process and the possible corrective actions needed in case of problems or opportunities. .

Chapter 6 presents a last simulation study that combines the results of the two previous chapters. More precisely, it validates and compares two alternative tracking methods and measures their efficiency on the total project objective. A top-down project tracking method relies on the EVM results of chapter 4 while a bottom-up tracking approach uses the results learnt from chapter 5.

Chapter 7 presents the new software tool ProTrack developed by OR-AS that integrates all research discussed throughout the various chapters in this book. Although the chapter does not enumerate all detailed features of the software, it gives an overview of the project scheduling and tracking approach and the different engines (project generation, simulation and time forecasting engines) that have been developed and discussed in this book.

Chapter 8 gives an overview of the various chapters presented throughout this book, and reviews the results from the four simulation studies from a project tracking point of view. More precisely, the conclusion clearly reviews the difference between top-down and bottom-up project tracking, and highlights the role of earned value management and schedule risk analysis in the two alternative tracking methods.

Most of the material and research has been published elsewhere. The work presented in chapter 1 can be found in the overview paper published by the International Journal of Project Management (Vandevoorde and Vanhoucke, 2006). Parts of the simulation study of chapter 4 have been published in the Measurable News (Vanhoucke and Vandevoorde, 2007a, 2008, 2009) and the Journal of the Operational Research Society (Vanhoucke and Vandevoorde, 2007b). Overview articles can be found in Vanhoucke (2008c,e, 2009). Other chapters or parts of chapters are still under submission (Vanhoucke, 2008a,b,d) and will hopefully be published soon in the academic literature.

Contents

List of Acronyms

A	
AC	Actual Cost
ACWP	Actual Cost of Work Performed
AD	Actual Duration
	Activity Distribution
ALAP	As-Late-As-Possible
AoA	Activity-on-the-Arc
AoN	Activity-on-the-Node
ASAP	As-Soon-As-Possible
AT	Actual Time
B	
BAC	Budget At Completion
BCWP	Budgeted Cost of Work Performed
BCWS	Budgeted Cost of Work Scheduled
BU-SRA	Bottom-up project tracking using Schedule Risk Analysis
C	
CI	Complexity Index
	Criticality Index
CNC	Coefficient of Network Complexity
CPM	Critical Path Method
CPI	Cost Performance Index
CPI(e)	Effective Cost Performance Index
CR	Critical Ratio
CRI	Cruciality Index
CV	Cost Variance
CV(e)	Effective Cost Variance
D	
d_i	Activity Duration

d_i^b	Activity duration (baseline schedule)
DFT	Due Finish Time
DSM	Design Structure Matrix
DST	Due Start Time

E

EAC	Expected At Completion (cost)
EAC(t)	Expected At Completion (time)
$EAC(t)_{PV}$	EAC(t) using the Planned Value method
$EAC(t)_{ED}$	EAC(t) using the Earned Duration method
$EAC(t)_{ES}$	EAC(t) using the Earned Schedule method
ED	Earned Duration
EDAC	Estimate of Duration At Completion
ES	Earned Schedule
EV	Earned Value
EVM	Earned Value Management
ESS	Earliest Start Schedule
E(x)	Expected value of variable x
ES(e)	Effective Earned Schedule
EV(e)	Effective Earned Value
EV(p)	Risk-free Earned Value
EV(r)	Risk-sensitive Earned Value

F

FF	Finish-Finish
FS	Finish-Start

G

GERT	Graphical Evaluation and Review Technique

I

IEAC(t)	Independent Estimate at Completion (time)
IEDAC	Independent Estimate of Duration At Completion

L

LA	Length of Arcs
LFT	Locked Finish Time
LRS	Latest Revised Schedule
LSS	Latest Start Schedule
LST	Locked Start Time

M

MAPE	Mean Absolute Percentage Error
MPE	Mean Percentage Error

N

N	Set of activities in a project network
n	Net of activities in a project network
n_l'	Number of arcs in a project network with length l
nrs	Number of simulation runs

O

OR-AS	Operations Research - Applications and Solutions
OS	Order Strength

P

PB	Project Buffer
PC	Percentage Completed
PCWR	Planned Cost of Work Remaining
PD	Planned Duration
PDWR	Planned Duration of Work Remaining
PERT	Programme Evaluation and Review Technique
PF	Performance Factor
P_i	Set of immediate predecessors of activity i
PLC	Project Life Cycle
PL_i	Progressive level of an activity i
PMBOK	Project Management Body Of Knowledge
PMI	Project Management Institute
ProTrack	Project Tracking (software tool)
PV	Planned Value
PV_{rate}	Planned Value Rate
PV_t	Planned Value at time t

R

RC	Remaining Cost
RD	Real Duration
RD^{no}	Real project duration without any corrective action
RD^{yes}	Real project duration with threshold triggered corrective action
RL_i	Regressive Level of activity i
RFT	Ready Finish Time
RST	Ready Start Time
R%	Estimated portion of EV(r) that is usable and requires no rework

S

SCI	Schedule Cost Index
SCI(t)	Schedule Cost Index (time)
SF	Start-Finish
S_i	Set of immediate successors of activity i

SI	Significance Index
SP	Serial/Parallel indicator
SPI	Schedule Performance Index
SPI(e)	Effective Schedule Performance Index
SPI(t)	Schedule Performance Index (time)
$\overline{\text{SPI(t)}}$	Average SPI(t) for all review periods
SPI(t)(e)	Effective Schedule Performance Index (time)
SPI to go	To Complete Schedule Performance Index
SRA	Schedule Risk Analysis
SS	Start-Start
SSI	Schedule Sensitivity Index
StDev	Standard Deviation
SV	Schedule Variance
SV(e)	Effective Schedule Variance
SV(t)	Schedule Variance (time)
SV(t)(e)	Effective Schedule Variance (time)
T	
TC	Total Contribution
TCSPI	To Complete Schedule Performance Index
TCSPI(t)	To Complete Schedule Performance Index (time)
$\text{TCSPI}_{\text{LRS}}$	To Complete Schedule Performance Index for LRS
$\text{TCSPI(t)}_{\text{LRS}}$	To Complete Schedule Performance Index (time) for LRS
TD-SPI	Top-down project tracking using SPI
TD-SPI(t)	Top-down project tracking using SPI(t)
TEAC	Time Estimate At Completion
TETC	Time Estimate To Complete
TF	Topological Float
tfi	Total Float of activity i
tf_i^k	Total Float of activity i at simulation run k
To complete SPI(t)	To Complete Schedule Performance Index for LRS
TV	Time Variance
U	
UC	Unit Contribution
UDR	Unearned Duration
W	
w_a	Width of a progressive level a
WBS	Work Breakdown Structure
Others	
σ_x	Standard Deviation of variable x
%C	Percentage Control
#◇	Number of times node (a) or node (b) of figure 6.2 is visited

List of Figures

List of Tables

Chapter 1
The EVM Fundamentals

Earned Value Management (EVM) is a methodology used since the 1960s, when the USA department of defense proposed a standard method to measure a project's performance. The system relies on a set of often straightforward metrics to measure and evaluate the general health of a project. These metrics serve as early warning signals to timely detect project problems or to exploit project opportunities. The purpose of an EVM system is to provide answers to project managers on questions such as:

- What is the difference between budgeted and actual costs?
- What is the current project status? Ahead of schedule or schedule delay?
- Given the current project performance, what is the expected remaining time and cost of the project?

Although EVM has been developed to measure and monitor both the time and cost dimension of a project, most attention has been unilaterally spent on the cost aspect of project management. Even the earned value guru's (Fleming and Koppelman, 2005) discuss the topic from a price tag point of view and stress in their well-known Harvard Business Review article (Fleming and Koppelman, 2003) that companies rely on some sort of EVM to predict the total project cost in a more accurate way than by simply using straightforward traditional cost accounting methods. However, the same authors (Fleming and Koppelman, 2004) openly ask the question why an EVM system, tailor-made to measure and monitor the performance of various projects, is rarely used in practice. They mention three important reasons why EVM has not been universally accepted on most projects, as follows:

- Language barrier: the terminology used in an EVM system does not belong to the daily language of the project manager and his/her team.
- The applicability: EVM was originally developed for major projects and not for the daily medium to small projects an average project manager is confronted with.
- The ostrich policy: management is often not interested in the real cost of a project in order to hide or postpone exuberant budget deviations.

M. Vanhoucke, *Measuring Time*, International Series in Operations Research & Management Science 136, DOI: 10.1007/978-1-4419-1014-1_1,

This chapter reviews the basic EVM metrics to measure the expected total project time and cost and to calculate the deviation between current and planned performance. The chapter discusses the recent renewed research attention on the time dimension of EVM and shows a number of anomalies and errors in the existing methods. This chapter has no intention to summarize all EVM related issues discussed throughout the literature, but serves as a fundament for the research studies presented in the following chapters.

1.1 Earned Value Management (EVM)

Earned Value Management is a methodology used to measure and communicate the real physical progress of a project and to integrate the three critical elements of project management (scope, time and cost management). It takes into account the work completed, the time taken and the costs incurred to complete the project and it helps to evaluate and control project risk by measuring project progress in monetary terms. The basic principles and the use in practice have been comprehensively described in many sources (for an overview, see e.g. Anbari (2003) or Fleming and Koppelman (2005)). Although EVM has been set up to follow up both time and cost, the majority of the research has been focused on the cost aspect (see e.g. the paper written by Fleming and Koppelman (2003) who discuss EVM from a price tag point of view). This chapter reviews the basic key metrics in earned value, elaborates on the recent research focused on the time aspect of EVM and compares a newly developed method, called earned schedule (Lipke, 2003), with the more traditional approach of forecasting a project's duration.

The outline of the chapter is as follows. In this section, the different metrics of an EVM system will be reviewed and will later be used in four simulation studies. Section 1.1.1 briefly reviews the EVM key parameters that serve as an input for the performance measures and the forecasting indicators (top layer of figure 1.1). Section 1.1.2 briefly reviews the existing performance measures (middle layer) and section 1.1.3 discusses the use of these performance measures to forecast the future performance of the project (bottom layer). Figure 1.1 serves as a guideline to sections 1.1.1, 1.1.2 and 1.1.3. All EVM metrics will be illustrated on a fictitious project network in section 1.2.

1.1.1 The metrics

Project performance should be measured throughout the life of the project and obviously requires a fixed time frame (i.e. a baseline schedule) for the project. A project schedule defines starting times (and finishing times) for each project activity and hence a planned value for each activity, both in terms of duration and costs. The planned duration PD equals the total project duration as a result of the constructed

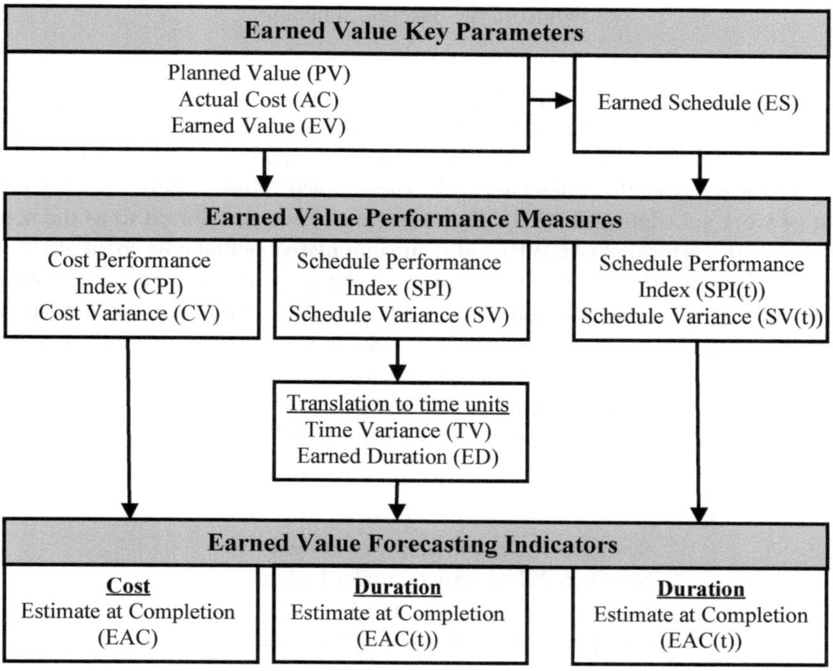

Fig. 1.1 Earned Value Management: key parameters, performance measures and forecasting indicators

CPM schedule and is often referred to as schedule at completion (SAC, Anbari (2003)). The actual time AT or actual duration AD defines the number of time periods (e.g. weeks) the project is in progress at the current time instance. Consequently, these measures are used to calculate the project progress and the number of time increments that the project is running, and are used to define the reporting periods for performance measurement from the start to the finish of the project. The real duration RD defines the real final project duration after execution. The budget at completion BAC is the sum of all budgeted costs for the individual activities. These variables can be summarized as follows:

PD	Planned Duration of the project → known from the baseline schedule = SAC (Schedule At Completion)
RD	Real Duration of the project → known at the finish of the project
AT	Actual Time in the project life → AT is a synonym for actual duration AD (AD = 1, ..., RD)
BAC	Budget At Completion

→ known from the baseline schedule

EVM requires three key parameters to measure project performance, i.e. the Planned Value (PV), the Actual Cost (AC) and the Earned Value (EV). The planned value is the time-phased budget baseline as an immediate result of the CPM schedule constructed from the project network. The planned value is often called budgeted cost of work scheduled (BCWS). The actual cost is often referred to as the actual cost of work performed (ACWP) and is the cumulative actual cost spent at a given point AT in time. The earned value represents the amount budgeted for performing the work that was accomplished by a given point AT in time. It is often called the budgeted cost of work performed (BCWP) and equals the total activity (or project) budget at completion multiplied by the percentage activity (or project) completion (PC) at the particular point in time (= PC * BAC). Figure 1.2 displays the three EVM key parameters for a fictitious project under the four different possible time/cost scenarios:

Scenario 1: late project, over budget
Scenario 2: late project, under budget
Scenario 3: early project, over budget
Scenario 4: early project, under budget

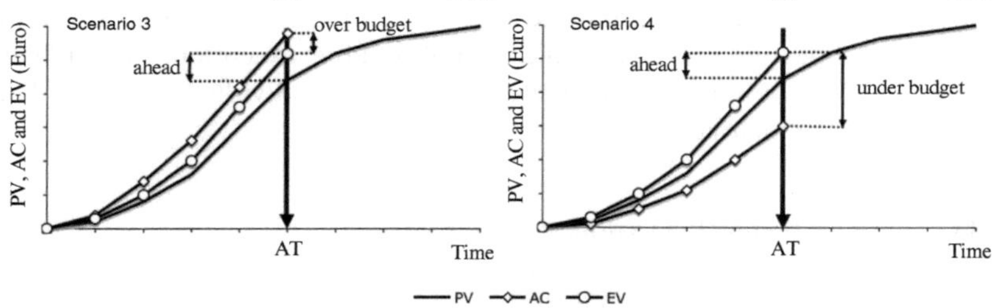

Fig. 1.2 The EVM key parameters PV, AC and EV for a project under 4 scenarios

1.1.2 Performance measures

Project performance, both in terms of time and costs, is determined by comparing the three key parameters PV, AC and EV, resulting in four well-known performance measures:

SV	Schedule Variance (SV = EV − PV)
SPI	Schedule Performance Index (SPI $= \frac{EV}{PV}$)
CV	Cost Variance (CV = EV − AC)
CPI	Cost Performance Index (CPI $= \frac{EV}{AC}$)

In the remainder of this book, these performance measures are calculated on the project level, and not on the level of each individual activity. Book (2006a,b), Jacob (2006) and Jacob and Kane (2004) criticize this approach and argue that the well-known performance measures are true indicators for project performance as long as they are used on the activity level, and not on the control account level or higher WBS (Work Breakdown Structure) levels. Jacob and Kane (2004) illustrate their statement with a simple example with two activities, leading to wrong and misleading results. As an example, a delay in a non-critical activity might give a warning signal that the project is in danger, while there is no problem at all since the activity only consumes part of its slack. Since the performance measures are calculated on the project level, this will lead to a false warning signal and hence, wrong corrective actions can be taken. It is generally recognized that effects of non-performing activities (delays) can be neutralized by well performing activities (ahead of schedule) at higher WBS levels, which might result in masking potential problems, but it is believed that this is the only approach that can be easily taken by practitioners. The earned value metrics are set up as early warning signals to detect in an easy and efficient way (i.e. at the cost account level, or even higher), rather than a simple replacement of the critical path based scheduling tools. This early warning signal, if analyzed properly, defines the need to eventually drill down into lower WBS levels. In conjunction with the project schedule, it allows taking corrective actions on those activities which are in trouble (especially those tasks which are on the critical path). As a result, the performance measures (SPI and SV) are calculated on the level of the project based on the three key indicators (PV, AC and EV) that are calculated per reporting period as the sum over all the individual activities (which can be easily done since they are expressed in monetary units). In chapter 4, a simulation study has been set up to measure the potential error of this project level based performance measuring approach on the accuracy of forecasting measuring to predict a project's final duration.

The cost performance indicators and their predictive power to forecast the final project cost (see next section) have been discussed extensively in literature and will not be repeated here. However, in order to track project time performance, the schedule performance measures need to be translated from monetary units to time units. In literature, three methods have been proposed to measure schedule performance: the planned value method (Anbari, 2003) and the earned duration method (Jacob and

Kane, 2004) translate the well-known SV and SPI indicators from monetary units to time units. The earned schedule method has been recently introduced by Lipke (2003) and calculates two alternative schedule performance measures (referred to as SV(t) and SPI(t)) that are directly expressed in time units.

The **planned value method** of Anbari (2003) relies on the well-known earned value metrics to forecast a project's duration using the following metrics:

PV_{rate}	Planned Value Rate (or planned accomplishment rate) $= \frac{BAC}{PD}$
TV	Time Variance $= \frac{SV}{PV_{rate}}$

The average planned value per time period, the planned value rate PV_{rate}, is defined as the baseline BAC divided by the planned duration PD. This measure can be used to translate the SV into time units, denoted by the time variance TV.

Jacob and Kane (2004) introduced a new term, **earned duration** ED, as the product of the actual duration and the SPI. Jacob (2003) and Jacob and Kane (2004) introduced the earned duration method as a reliable methodology for forecasting a project's final duration using the schedule performance index SPI.

ED	Earned duration $= AD * SPI$

The earned duration ED is the product of the actual duration AD and the schedule performance index SPI, and translates the current actual project duration AD into an earned duration ED taking the current schedule performance into account. Consequently, projects with a delay (i.e. $SPI < 1$) have an earned duration ED lower than the current actual duration while well performing projects (i.e. $SPI > 1$) have earned more time than actually needed, i.e. $ED > AD$.

Lipke (2003) criticized the use of the classic SV and SPI metrics since they give false and unreliable time forecasts near the end of the project. Instead, he provided a time-based measure to overcome the quirky behavior of the SV and SPI indicators. This **earned schedule method** relies on similar principles of the earned value method, and uses the concept of earned schedule (ES) as follows:

Find t such that $EV \geq PV_t$ and $EV < PV_{t+1}$

$$ES = t + \frac{EV - PV_t}{PV_{t+1} - PV_t}$$

with

ES	Earned Schedule
EV	Earned Value at the actual time
PV_t	Planned Value at time instance t

The cumulative value for the ES is found by using the EV to identify in which time increment t of PV the cost value for EV occurs. ES is then equal to the cumulative time t to the beginning of that time increment, plus a fraction $\frac{EV-PV_t}{PV_{t+1}-PV_t}$ of it. The fraction equals the portion of EV extending into the incomplete time increment divided by the total PV planned for that same time period, which is simply calculated as a linear interpolation between the time-span of time increment t and t + 1. Note that the formula description is not completely mathematically correct in case $EV = PV_t = PV_{t+1}$. In this case, the ES is equal to the earliest period t for which EV $= PV_t$. This is, for example, the case at the end of the project, where the ES metric is equal to the PD, and EV = BAC. Figure 1.3 shows a graphical fictitious example of the linear interpolation of the planned values between review period t and t+1.

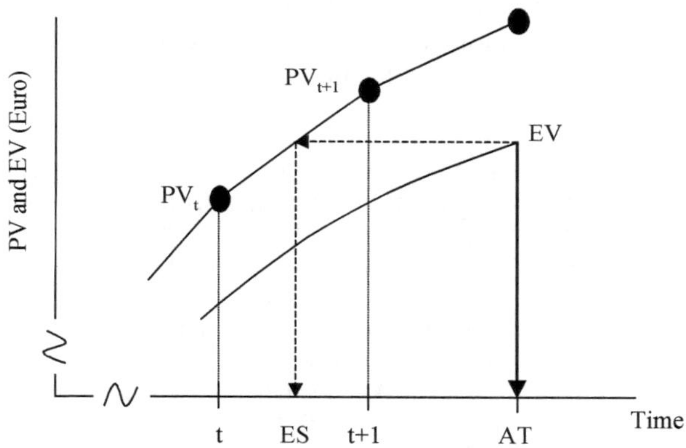

Fig. 1.3 Linear interpollation between PV_t and PV_{t+1}

Figure 1.4 illustrates the translation of the earned value into the ES metric to clearly show whether a project is behind (left) or ahead of (right) schedule.

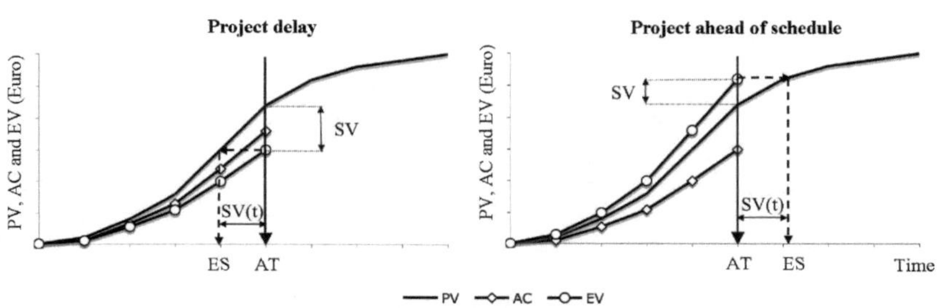

Fig. 1.4 The ES metric for a late (left) and early (right) project

Using the ES concept, two indicators can be constructed which serve as good and reliable alternatives of the SV and SPI indicators, as follows:

SV(t)	Schedule Variance with earned schedule ES – AT
SPI(t)	Schedule Performance Index with earned schedule ES / AT

Table 1.1 and figure 1.5 clearly display the unreliable behavior of the SV and SPI metrics for a project that finishes later than planned (PD = 9 weeks while the real duration RD = 12 weeks). The last review periods of the project are unreliable since both the SV and SPI metrics clearly show an improving trend. At the end of the project, both metrics give a signal that the project finishes within time (SV = 0 and SPI = 1 at the end of the project), although it is 3 weeks late. The SV(t) and SPI(t) metrics give a correct signal along the whole life of the project. The SV(t) equals -3 at the end of the project, which is a reflection of the 3 weeks delay.

Table 1.1 Numerical example: The SV/SPI versus SV(t)/SPI(t) performance measures

AT	PV	EV	ES	SV	SPI	SV(t)	SPI(t)
0	0.00	0.00					
1	1.00	0.80	0.80	-0.20	0.80	-0.20	0.80
2	4.00	2.75	1.58	-1.25	0.69	-0.42	0.79
3	8.00	5.50	2.38	-2.50	0.69	-0.63	0.79
4	15.00	10.00	3.29	-5.00	0.67	-0.71	0.82
5	22.00	15.00	4.00	-7.00	0.68	-1.00	0.80
6	26.00	19.00	4.57	-7.00	0.73	-1.43	0.76
7	28.00	23.00	5.25	-5.00	0.82	-1.75	0.75
8	29.00	25.00	5.75	-4.00	0.86	-2.25	0.72
9	30.00	27.00	6.50	-3.00	0.90	-2.50	0.72
10	30.00	28.00	7.00	-2.00	0.93	-3.00	0.70
11	30.00	29.00	8.00	-1.00	0.97	-3.00	0.73
12	30.00	30.00	9.00	0.00	1.00	-3.00	0.75

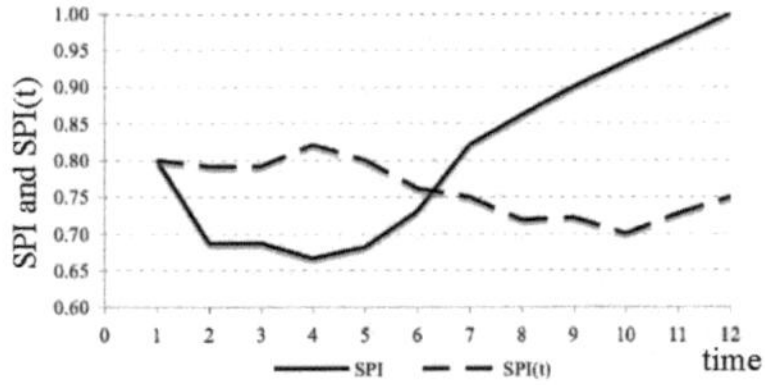

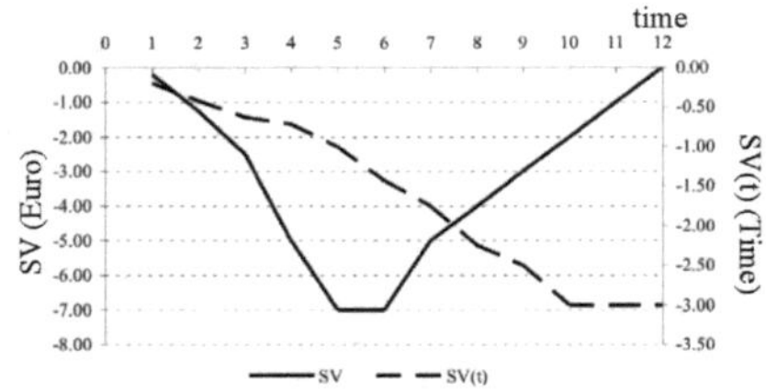

Fig. 1.5 The SPI and SV versus SPI(t) and SV(t) performance measures

Since the introduction of the earned schedule concept by Lipke (2003), other authors have investigated the potential of the new method in various ways. Henderson (2003) has shown the validity of the ES concepts on a portfolio of six projects. In another paper, he extended this novel approach (Henderson, 2004) and used it on a small scale but time critical information technology software development project (Henderson, 2005). Hecht (2007) used data from a U.S. Navy project to build a helicopter trainer for maintenance personnel and used it as a case study to test the predictive power of the earned schedule method. Henderson and Zwikael (2008) give a summary of their project performance stability study by investigating a large set of projects from three different countries. Vandevoorde and Vanhoucke (2006) were the first authors that extensively compared the three methods and tested them to a simple one activity project and a real life data set. They summarized the often confusing terminology used in the earned value/schedule literature. Lipke et al (2008) have statistically validated the earned schedule method based on a pool of real life project data. The current chapter of this book is a summary of the work presented by Vandevoorde and Vanhoucke (2006).

In the remainder of this chapter, an additional performance index will be used, known as the Schedule Cost Index, and defined as:

SCI	Schedule Cost Index (using the traditional SPI) SPI * SCI
SCI(t)	Schedule Cost Index (using the SPI(t)) SPI(t) * SCI

1.1.3 Forecasting formula

One of the primarily tasks of a project manager is making decisions about the future. EVM systems are designed to follow up the performance of a project and to act as a warning signal to take corrective actions in the (near) future. Forecasting the total project cost and the time to completion is crucial to take corrective actions when problems or opportunities arise and hence, the performance measures will be mainly used as early warning signals to detect these project problems and/or opportunities. EVM metrics are designed to forecast these two important performance measures (time and cost) based on the actual performance up to date and the assumptions about future performance. In this section, some generally accepted and newly developed forecasting measures will be reviewed and will be used throughout all chapters of this book.

The general formula for predicting a project's final cost is given by the Estimated cost At Completion (EAC), as follows:

EAC = AC + PCWR

with

EAC	Estimated cost at Completion
AC	Actual Cost
PCWR	Planned Cost of Work Remaining

The general and similar formula for predicting a project's total duration is given by the Estimated duration At Completion (EAC(t)), as follows:

EAC(t) = AD + PDWR

with

EAC(t)	Estimated duration at Completion
AD	Actual Duration (or Actual Time AT)
PDWR	Planned Duration of Work Remaining

Note that the abbreviation EAC is used for cost forecasting and at between brackets is added (i.e. EAC(t)) for time forecasting. Cost performance and forecasting have been widely investigated by numerous researchers, and is outside the scope of this chapter. For an overview, the reader is referred to Christensen (1993) who reviews different EAC formulas and several studies that examine their accuracy. Figure 1.6 shows a fictitious project with estimated values for the final project duration EAC(t) (the overrun EAC(t) - PD is often referred to as the project slippage) and the estimated final cost overrun EAC.

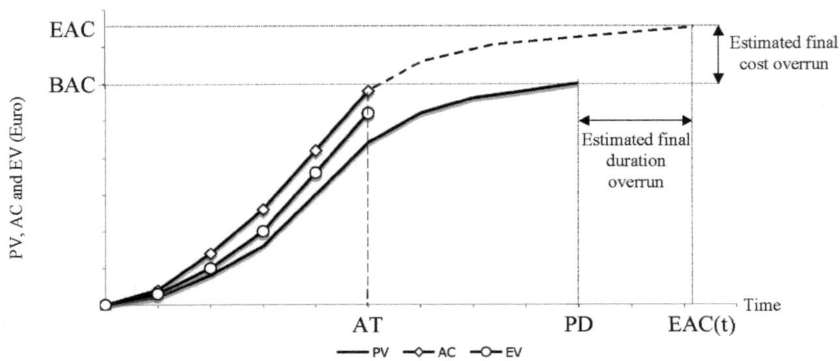

Fig. 1.6 Expected cost and time performance

The remainder of this chapter compares and validates the different methods to forecast a project's final duration. Note that EAC(t) is often referred to as the Time

Estimate at Completion (TEAC), the Estimate of Duration at Completion (EDAC) or the Independent Estimate of Duration at Completion (IEDAC). The terminology used in the current chapter is based on the terminology summarized by Vandevoorde and Vanhoucke (2006) who compared three methods to estimate the PDWR based on research done by Anbari (2003), Jacob (2003) and Lipke (2003). Each method has three different versions to predict a project's final duration, depending on the characteristics and performance of the project in the past. Table 1.2 summarizes the forecasting metrics used in this book. The PDWR metric is the component that has to be estimated, and heavily depends on the specific characteristics and the current status of the project (Anbari, 2003). The table makes a distinction between six different project situations based on the classification described in Anbari (2003).

The first project situation assumes ideal circumstances and does not require any forecasting since the project is considered to be on plan. The second and third row refer to project situations where forecasting (i.e. estimating the PDWR) is useless due to the changing conditions or irreversible problems. Hence, the remainder of this book will focus on the last three possible project scenarios. In these cases, it is assumed that the remaining work PDWR will be done according to plan (scenario 4), will follow the current SPI trend (scenario 5) or will follow the current SCI trend (scenario 6). Each forecasting technique described in the three following subsections will be discussed from these last three project scenarios point of view.

Only three project duration forecasting methods have been presented in literature, referred to as the planned value method (Anbari, 2003), the earned duration method (Jacob, 2003) and the earned schedule method (Lipke, 2003). However, the many notations, abbreviations and often confusing metrics used to describe these three methods unnecessarily complicate the comparability of these methods. In order to shed light on the confusing terminology, the overwhelming amount of synonyms taken from these various literature sources has been summarized in table 1.3[1]. The table illustrates the confusing terminology for the three forecasting methods used throughout the literature. The row labelled with "duration measures" displays the terminology used to refer to the general duration forecasting formula EAC(t) = AD + PDWR. The row labelled with "assessment indicators" displays the terminology used to measure the additional effort needed to finish the project within the project deadline. The specific calculation of these metrics will be explained in detail in the following three subsections. Throughout the book, a more standardized terminology will be used to avoid confusion between the different methods. Table 1.4 displays the terminology used and can be considered as a standardization of the terminology of table 1.3.

[1] The terminology used is based on the presentation by Lipke and Henderson "Earned schedule - an emerging practice" presented at the 16th Annual International Integrated Program Management Conference, November 15-17, 2004, Virginia.

Table 1.2 The estimated PDWR depending on the project situation (Source: Vandevoorde and Vanhoucke (2006), and based on Anbari (2003))

	Forecasting method			
Situation	**Anbari (2003)**	**Jacob (2003)**	**Lipke (2003)**	**Comments**
EAC(t) as originally planned		monitor schedule		The final project duration will be as planned, regardless of the past performance. This situation may be dangerous, as unattended problems mostly do not resolve themselves (well catch up during the commissioning phase)
PDWR is new		re-schedule		The original project assumptions are no longer valid for the remaining work (due to changed conditions). The use of performance indices to predict is obsolete and a new schedule for the remaining work needs to be developed
PDWR is very high		re-schedule		Quality problems are irreversible and a lot of extra time is needed to fix the problems (occurs mostly in the late project stage). Stakeholders usually lose their interest in the project ("If this project ever finishes, it would be a miracle")
PDWR according to plan	$EAC(t)_{PV1}$	$EAC(t)_{ED1}$	$EAC(t)_{ES1}$	Past performance is not a good predictor of future performance. Problems/opportunities of the past will not affect the future, and the remaining work will be done according to plan
PDWR will follow current SPI trend	$EAC(t)_{PV2}$	$EAC(t)_{ED2}$	$EAC(t)_{ES2}$	Past schedule performance is a good predictor of future performance. Problems/opportunities of the past will affect future performance, and the remaining work will be corrected for the observed efficiencies or inefficiencies
PDWR will follow current SCI trend	$EAC(t)_{PV3}$	$EAC(t)_{ED3}$	$EAC(t)_{ES3}$	Past cost and schedule problems are good indicators for future performance (i.e. cost and schedule management are inseparable). The SCI = SPI * CPI (Schedule Cost Index) is often called the critical ratio index
	Planned Value	**Earned Duration**	**Earned Schedule**	

Table 1.3 Terminology used in comparison papers

	Planned value method Anbari (2003)		**Earned duration method Jacob (2003)**		**Earned schedule method Lipke (2003)**	
Baseline	SAC	Schedule at Completion	PD	Planned Duration	PD	Planned Duration
Status of the project (duration)	PV_{rate}	Planned Value Rate (BAC/PD)	ED	Earned Duration (AD/SPI)	ES	Earned Schedule
	AT	Actual Time	AD	Actual Duration	AT	Actual Time
	SPI	Schedule Performance Index (EV/PV)	SPI	Schedule Performance Index (EV/PV)	SPI(t)	Schedule Performance Index Time (ES/AT)
	SV	Schedule Variance	SV	Schedule Variance (EV-PV)	SV(t)	Schedule Variance Time (ES-AT)
	SV	(EV-PV)				
	TV	Time Variance (SV/PV_{rate})	—	—	—	—
	CR	Critical Ratio	—	—	SCI(t)	Schedule Cost Index
Duration measures	**TEAC = AT + TETC**		**EDAC = AD + UDR**		**EAC(t) = AT + PDWR**	
	TETC	Time Estimate to Complete	UDR	Unearned Duration Remaining	PDWR	Planned Duration for Work Remaining
	TEAC	Time Estimate at Completion	EDAC	Estimate of Duration at Completion	EAC(t)	Estimate of Duration at Completion
	—	—	—	—	**IEAC(t) = AT + PDWR / PF**	
	—	—	—	—	IEAC(t)	Independent Estimate at Completion Time
Assessment indicators	—	—	TCSPI	To Complete Schedule Performance Index	SPI(t) to go[a]	To Complete Schedule Performance Index for PD
	—	—	—	—	To Complete SPI(t)[b]	To Complete Schedule Performance Index for Latest Revised Schedule (LRS)

[a] The SPI(t) to go is equal to the TCSPI or the TCSPI(t) of this chapter.

[b] The to complete SPI(t) equals the $TCSPI_{LRS}$ or the $TCSPI(t)_{LRS}$ of this chapter.

1.1.3.1 The planned value method

The planned value method described by Anbari (2003) does not directly give an estimate for the PDWR but relies on the planned value rate which is equal to the average planned value per time period, i.e. $PV_{rate} = \frac{BAC}{PD}$ where BAC is used to denote the budget at completion and PD to denote total planned project duration. This method assumes that the schedule variance can be translated into time units by dividing the schedule variance by the planned value rate, resulting in the time variance TV as follows:

$$TV = \frac{SV}{PV_{Rate}} = \frac{SV * PD}{BAC} = \frac{(EV - PV) * PD}{BAC} \tag{1.1}$$

According to the project characteristics (reflected by the last three situations of table 1.2), the following forecasting formulas have been derived:

- EAC(t) with the duration of remaining work as planned

  $$EAC(t)_{PV1} = PD - TV$$

- EAC(t) with the duration of remaining work following the current SPI trend

  $$EAC(t)_{PV2} = \frac{PD}{SPI}$$

- EAC(t) with the duration of remaining work following the current SCI trend

  $$EAC(t)_{PV3} = \frac{PD}{SCI}$$

Note that the terminology of Anbari (2003) is somewhat different since he proposes the "Time Estimate at Completion (TEAC)" and the "Time Estimate to Complete (TETC)" to refer to the EAC(t) and the PDWR, respectively (see table 1.3).

1.1.3.2 The earned duration method

The earned duration method is described by Jacob (2003) and extended by Jacob and Kane (2004). The earned duration ED is the product of the actual duration AD and the schedule performance index SPI, i.e. ED = AD * SPI, and hence, the generic earned duration forecasting formula is:

$$EAC(t)_{ED} = AD + \frac{PD - ED}{PF} \tag{1.2}$$

The performance factor PF is used to adapt the future performance to the past performance (depending on the project characteristics) and reflects the last three situations of table 1.2. Indeed, a PF = 1 denotes a future performance at the efficiency rate of the original plan (100% efficiency). However, the future performance can be corrected towards the current SPI trend or the current SCI = CPI * SPI trend. Hence, the three forecasting methods to predict total project duration are:

- PF = 1: Duration of remaining work as planned

$$\mathrm{EAC(t)_{ED1}} = \mathrm{AD} + (\mathrm{PD} - \mathrm{ED}) = \mathrm{PD} + \mathrm{AD} * (1 - \mathrm{SPI})$$

- PF = SPI: Duration of remaining work with SPI trend

$$\mathrm{EAC(t)_{ED2}} = \mathrm{AD} + \frac{\mathrm{PD} - \mathrm{ED}}{\mathrm{SPI}} = \frac{\mathrm{PD}}{\mathrm{SPI}}$$

- PF = SCI: Duration of remaining work with SCI trend[2]

$$\mathrm{EAC(t)_{ED3}} = \mathrm{AD} + \frac{\mathrm{PD} - \mathrm{ED}}{\mathrm{SCI}} = \frac{\mathrm{PD}}{\mathrm{SCI}} + \mathrm{AD} * (1 - \frac{1}{\mathrm{CPI}})$$

In situations where the actual project duration exceeds the planned duration (i.e. AD > PD), and the work is not yet completed, the PD will be substituted by the AD in the above mentioned formulas. In these cases, the formulas are:

$$\mathrm{EAC(t)_{ED1}} = \mathrm{AD} + (\mathrm{AD} - \mathrm{ED}) = \mathrm{AD} * (2 - \mathrm{SPI})$$

$$\mathrm{EAC(t)_{ED2}} = \mathrm{AD} + \frac{\mathrm{AD} - \mathrm{ED}}{\mathrm{SPI}} = \frac{\mathrm{AD}}{\mathrm{SPI}}$$

$$\mathrm{EAC(t)_{ED3}} = \mathrm{AD} + \frac{\mathrm{AD} - \mathrm{ED}}{\mathrm{SCI}} = \mathrm{AD} * (1 - \frac{1}{\mathrm{CPI}} + \frac{1}{\mathrm{SCI}})$$

An additional assessment metric given by Jacob (2003) measures the additional effort needed to finish the project within the project deadline. This corrective action metric related to the schedule performance is called the "To Complete Schedule Performance Index" (TCSPI) and is calculated as:

$$\mathrm{TCSPI} = \frac{\mathrm{PD} - \mathrm{ED}}{\mathrm{PD} - \mathrm{AD}}$$

or

$$\mathrm{TCSPI_{LRS}} = \frac{\mathrm{PD} - \mathrm{ED}}{\mathrm{LRS} - \mathrm{AD}}$$

[2] This is an additional forecasting formula which is not reported in the original manuscript of Jacob (2003).

The former measures the additional effort needed to finish the project within the original planned duration while the latter measures the effort to finish the project with the latest revised schedule (LRS) duration, which is an updated planned duration after an intermediate corrective action taken during the project life time.

1.1.3.3 The earned schedule method

The generic earned schedule duration forecasting formula is:

$$\mathrm{EAC(t)_{ES}} = \mathrm{AD} + \frac{\mathrm{PD} - \mathrm{ES}}{\mathrm{PF}} \tag{1.3}$$

The performance factor used depends on the project situation:

- PF = 1: Duration of remaining work as planned

$$\mathrm{EAC(t)_{ES1}} = \mathrm{AD} + (\mathrm{PD} - \mathrm{ES})$$

- PF = SPI(t): Duration of remaining work with SPI(t) trend

$$\mathrm{EAC(t)_{ES2}} = \mathrm{AD} + \frac{\mathrm{PD} - \mathrm{ES}}{\mathrm{SPI(t)}}$$

- PF = SCI(t): Duration of remaining work with SCI(t) trend (note that this formula is not given in any of the earned schedule papers)

$$\mathrm{EAC(t)_{ES3}} = \mathrm{AD} + \frac{\mathrm{PD} - \mathrm{ES}}{\mathrm{CPI} * \mathrm{SPI(t)}} = \mathrm{AD} + \frac{\mathrm{PD} - \mathrm{ES}}{\mathrm{SCI(t)}}$$

The "To Complete Schedule Performance Index" or TCSPI(t) can be calculated as:

$$\mathrm{TCSPI(t)} = \frac{\mathrm{PD} - \mathrm{ES}}{\mathrm{PD} - \mathrm{AD}}$$

or

$$\mathrm{TCSPI(t)_{LRS}} = \frac{\mathrm{PD} - \mathrm{ES}}{\mathrm{LRS} - \mathrm{AD}}$$

and measures the additional effort to finish the project within the planned duration or the revised duration, respectively.

Remark that the TCSPI(t) and the $\mathrm{TCSPI(t)_{LRS}}$ is denoted as the "SPI(t) to go" and the "to complete SPI(t)" in table 1.3. Table 1.4 displays the terminology used throughout this book and can be considered as an update and standardization of the often confusing terminology used in various sources in literature, as summarized in table 1.3.

Table 1.4 Terminology used throughout the book

	Planned value method Anbari (2003)		**Earned duration method Jacob (2003)**		**Earned schedule method Lipke (2003)**	
Baseline	SAC	Schedule at Completion	PD	Planned Duration	PD	Planned Duration
Status of the project (duration)	PV_{rate}	Planned Value Rate (BAC/PD)	ED	Earned Duration (AD/SPI)	ES	Earned Schedule
	AT	Actual Time	AD	Actual Duration	AT	Actual Time
	SPI	Schedule Performance Index (EV/PV)	SPI	Schedule Performance Index (EV/PV)	SPI(t)	Schedule Performance Index Time (ES/AT)
	SV	Schedule Variance	SV	Schedule Variance (EV-PV)	SV(t)	Schedule Variance Time (ES-AT)
	SV	(EV-PV)				
	TV	Time Variance (SV/PV_{rate})	—	—	—	—
	CR	Critical Ratio	—	—	SCI(t)	Schedule Cost Index
Duration measures	**EAC(t) = AD + PDWR**		**EAC(t) = AD + (PD - ED) / PF**		**AD + (PD - ES) / PF**	
	$EAC(t)_{PV1}$	Estimate of Duration at Completion PF = 1	$EAC(t)_{ED1}$	Estimate of Duration at Completion PF = 1	$EAC(t)_{ES1}$	Estimate of Duration at Completion PF = 1
	$EAC(t)_{PV2}$	Estimate of Duration at Completion PF = SPI	$EAC(t)_{ED2}$	Estimate of Duration at Completion PF = SPI	$EAC(t)_{ES2}$	Estimate of Duration at Completion PF = SPI
	$EAC(t)_{PV3}$	Estimate of Duration at Completion PF = SCI	$EAC(t)_{ED3}$[a]	Estimate of Duration at Completion PF = SCI	$EAC(t)_{ES3}$[b]	Estimate of Duration at Completion PF = SCI
Assessment indicators	—	—	TCSPI	To Complete Schedule Performance Index	TCSPI(t)	To Complete Schedule Performance Index for PD
	—	—	—	—	$TCSPI(t)_{LRS}$	To Complete Schedule Performance Index for Latest Revised Schedule (LRS)

[a] This forecasting formula does not appear in the original paper by Jacob (2003).

[b] This forecasting formula does not appear in the original paper by Lipke (2003).

1.2 A fictitious project example

Consider a project network example in figure 1.7 that will be used throughout the various chapters. It is assumed that a project is represented by an activity-on-the-node network G = (N, A) where the set of nodes, N, represents network activities and the set of arcs, A, represents the technological precedence relations between the activities[3].

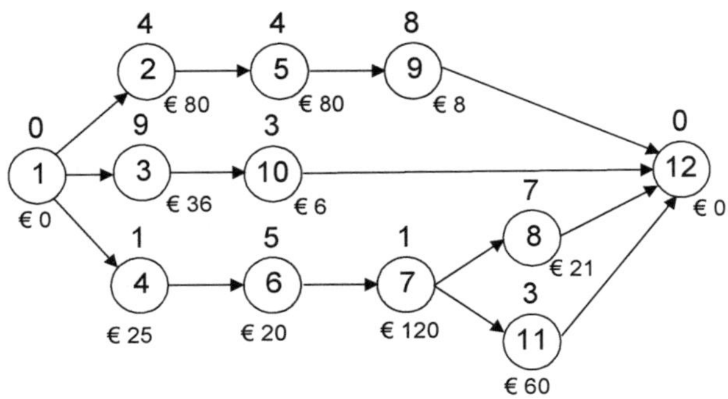

Fig. 1.7 An example project

Each activity has an estimated duration as denoted above each node in figure 1.7, with a corresponding budgeted cost denoted below the node. Activities 1 and 12 are dummies and are used to denote the start and end of the project (with a zero duration and a zero cost). A schedule is defined by a vector of starting times which implies a vector of finishing times, as given in the Gantt chart of figure 1.8. This figure displays the baseline earliest start schedule with a total planned duration PD equal to 16. The budget at completion is the sum of the individual activity costs which equals BAC = € 456.

In order to show the detailed calculations of all forecasting measures, it is assumed that the project execution is known and finishes with a two-periods delay as displayed in figure 1.9. This figure displays the (fictitious) real life execution (actual values for each activity duration and the corresponding actual costs) of all project activities with a real project duration RD equal to 18. The final (real) total project cost amounts to € 643, which is higher than the BAC value due to the activity delays (see table 1.6 for detailed calculations). The project tracking process assumes a reporting period per time unit, from AD = 1, 2, ..., 18.

Table 1.6 reports the cumulative values for the three earned value key parameters on the level of an individual activity as well as the performance measures on

---

[3] Alternatively, one could model a project network as an activity-on-the-arc network where arcs represent the project activities and nodes represent project events to implicitly model the precedence constraints. Since most commercial software use the AoN representation, this alternative project network representation is not used in this book.

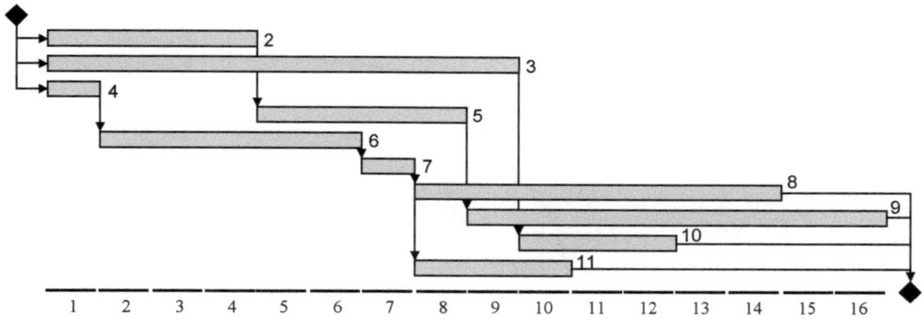

Fig. 1.8 The baseline schedule for the project of figure 1.7

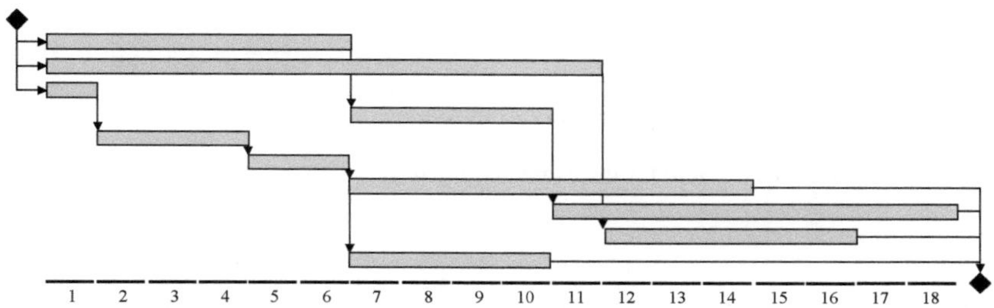

Fig. 1.9 The actual project execution Gantt chart of the example project

the level of the project along the life of the example project. The top displays the planned value key parameter for each project activity for each period of the baseline schedule (i.e. PD = 1, 2, ..., 16). As an example, the cumulative planned value for activity 2 increases with $\frac{80}{4}$ = € 20 per period. The second and the third part of the table displays the cumulative values for the two remaining key parameters for each project activity along the life of the project (i.e. AT = 1, 2, ..., 18). The actual cost for activity 2 amounts to € 120 in total due to a two days delay. The cumulative earned value measures the amount that is earned per period during its execution time, and hence, equals $\frac{80}{6}$ = € 13.33 per period. The bottom of the table displays the performance metrics of the different methods described above, calculated on the level of the project. Figure 1.10 displays the traditional S-curve of the PV, EV and AC project metrics along the time horizon of the example project. This figure reveals a temporary project ahead of schedule situation at time instances 5, 6 and 7 (EV $>$ PV), followed by a project delay (EV $<$ PV) resulting in the final two days delay. This situation is reflected by the earned schedule metric of table 1.6 which clearly shows that ES $>$ AT at time instances 5 to 7 and ES $<$ AT the periods afterwards. As an example, the ES at time period 6, denoted by ES_6, equals $6 + \frac{264.64-189}{333-189} = 6.53$, since EV $\geq PV_6$ (264.64 $>$ 189) and EV $< PV_7$ (264.64 $<$ 333).

Figure 1.11 shows the forecasting values for each review period (from 1 to 18), grouped according to the performance factor PF (see tables 1.2 and 1.4). The top,

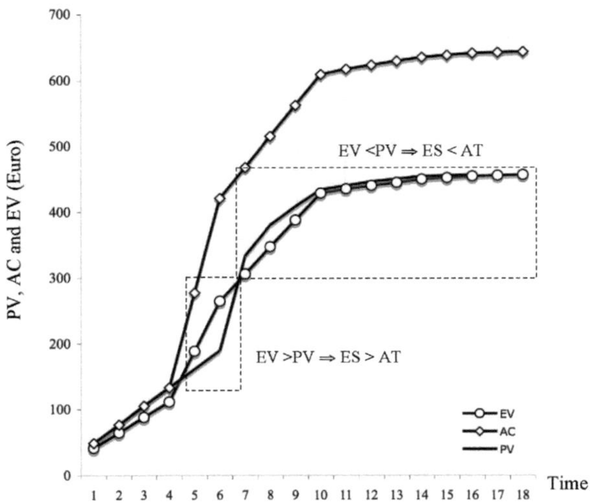

Fig. 1.10 The traditional S-curve for the example project

middle and bottom graphs display the forecasts with a performance factor PF = 1, PF = SPI and PF = SPI*CPI, respectively. The figures reveal that all forecasting methods report an underestimation for the final project duration when PF = 1, while both under- and overestimates occur for the PF = SPI and PF = SPI*SPI forecasting methods. Obviously, these results are case-specific and only hold for the fictitious project example and hence, they cannot be generalized to any project setting.

The purpose of chapter 4 is to test the accuracy of these duration forecasting methods on a wide and diverse set of projects in order to avoid case-specific conclusions and to generalize results to a broader project setting. As an example, table 1.5 displays a measure for the forecast accuracy[4] as the absolute percentage deviation between the actual duration forecast EAC(t) and the real duration RD = 18. A forecasting accuracy value is shown for each review period (from 1 to 18, see first column) while the average forecasting accuracy is displayed in the last row. The table reveals that the $\mathrm{EAC(t)}_{\mathrm{ES1}}$ and $\mathrm{EAC(t)}_{\mathrm{ES2}}$ forecasting metrics have the best average forecasting accuracy. As an example, the forecast $\mathrm{EAC(t)}_{\mathrm{ES2}}$ of period 6 is equal to $6+\frac{16-6.53}{1.09}=14.69$ (i.e. an underestimation of the final project duration) and consequently, the accuracy equals $\frac{|18-14.69|}{18}*100=18.40\%$. A detailed study of the forecast accuracy on a diverse set of projects is the subject of chapter 4.

[4] All values have been calculated based on the rounded values reported in table 1.6. Consequently, small rounding errors might occur.

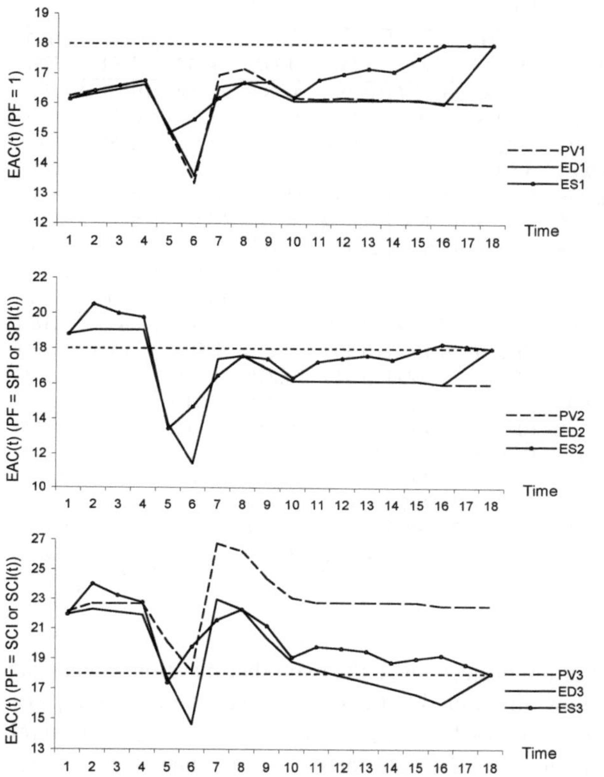

Fig. 1.11 The 9 duration forecasts along the life of the project

1.3 Conclusion

This chapter briefly reviewed the basic key parameters and performance measures of earned value management systems and discussed three different project duration methods that can be used based on these earned value parameters. A generic formula to forecast the total duration of a project has been presented, and has been further split into different project situations. More precisely, each method (planned value method, earned duration method and earned schedule method) can be further subdivided into three different forecasting submodels as a function of the current project situation and the assumption about future expected performance.

In this chapter, it is conjectured that the use of the planned value method, the earned duration method or the earned schedule method might lead to similar results for project monitoring in the early and middle stages. However, it might be recommended to shift to the earned schedule method for monitoring project progress at

Table 1.5 The forecasting accuracy along the life of the example project

	EAC(t)			EAC(t)			EAC(t)		
	PV1	**PV2**	**PV3**	**ED1**	**ED2**	**ED3**	**ES1**	**ES2**	**ES3**
1	9.67%	4.58%	23.03%	10.28%	4.58%	22.05%	10.28%	4.58%	22.05%
2	8.75%	5.82%	25.98%	9.33%	5.82%	23.86%	8.72%	13.89%	33.47%
3	7.83%	5.82%	25.98%	8.44%	5.82%	22.80%	7.78%	11.11%	29.10%
4	6.90%	5.82%	25.98%	7.56%	5.82%	21.74%	6.83%	9.81%	26.49%
5	16.38%	24.03%	11.73%	15.83%	24.03%	1.35%	16.50%	25.40%	3.36%
6	25.86%	36.51%	0.78%	24.44%	36.51%	18.80%	14.06%	18.40%	9.95%
7	5.76%	3.38%	48.64%	8.00%	3.38%	27.70%	10.06%	8.48%	19.87%
8	4.57%	2.32%	45.79%	7.11%	2.32%	23.90%	7.17%	2.38%	23.81%
9	7.08%	6.43%	35.60%	8.61%	6.43%	13.14%	7.00%	3.26%	17.74%
10	9.99%	10.21%	28.27%	10.56%	10.21%	4.46%	9.89%	9.18%	5.93%
11	10.16%	10.21%	26.46%	10.50%	10.21%	1.50%	6.61%	4.18%	9.99%
12	9.93%	10.21%	26.46%	10.44%	10.21%	0.77%	5.50%	3.08%	9.28%
13	10.09%	10.21%	26.46%	10.39%	10.21%	3.04%	4.44%	2.14%	8.34%
14	10.25%	10.21%	26.46%	10.33%	10.21%	5.31%	5.00%	3.50%	4.14%
15	10.49%	10.21%	26.46%	10.28%	10.21%	7.58%	2.50%	0.93%	5.50%
16	10.72%	11.11%	25.20%	11.11%	11.11%	11.11%	0.00%	1.52%	6.67%
17	10.92%	11.11%	25.20%	5.56%	5.56%	5.56%	0.00%	0.76%	3.34%
18	11.11%	11.11%	25.20%	0.00%	0.00%	0.00%	0.00%	0.00%	0.00%
Avg.	10.36%	10.52%	26.65%	9.93%	9.59%	11.93%	6.80%	6.81%	13.28%

the final stage of the project, since the two other duration forecasting methods have been said to show an unreliable trend at the late stages of the project. The earned schedule, on the contrary, has been developed to overcome this unreliable behavior, possibly leading to more accurate forecasts at the late stages of the project. Moreover, it is also recommended to use these schedule forecasting methods at least at the cost account level or at higher levels of the work breakdown structure. This is contradictory to the statements given by Jacob (2003) who argues that the schedule forecast metrics should only be used at the level of the activity. This concern has also been raised by other authors and has led to a discussion summarized in articles such as Book (2006a,b), Jacob (2006) and Lipke (2006). Although it is recognized that, at higher WBS levels, effects (delays) of non-performing activities can be neutralized by well performing activities (ahead of schedule), which might result in masking potential problems, it is the only approach that can be taken by practitioners. Indeed, the earned value metrics are set up as early warning signals to detect problems and/or opportunities in an easy and efficient way (i.e. at the cost account level, or even higher), rather than a simple replacement of the critical path based scheduling tools. This early warning signal, if analyzed properly, defines the need to eventually drill down into lower WBS levels. In conjunction with the project schedule, it allows to take corrective actions on those activities which are in trouble (especially those tasks which are on the critical path). Lipke et al (2008) also note that detailed schedule analysis is a burdensome activity and if performed often can have disrupting effects on the project team. EVM offers calculation methods yielding reliable

Table 1.6 The cumulative planned value PV, actual cost AC and earned value EV for each activity along the life of the example project and the performance measures on the project level

PV	1	2	3	4	5	6	7	8	9	10	11	12	13	14	15	16		
2	20	40	60	80	80	80	80	80	80	80	80	80	80	80	80	80	-	-
3	4	8	12	16	20	24	28	32	36	36	36	36	36	36	36	36	-	-
4	25	25	25	25	25	25	25	25	25	25	25	25	25	25	25	25	-	-
5	0	0	0	0	20	40	60	80	80	80	80	80	80	80	80	80	-	-
6	0	4	8	12	16	20	20	20	20	20	20	20	20	20	20	20	-	-
7	0	0	0	0	0	0	120	120	120	120	120	120	120	120	120	120	-	-
8	0	0	0	0	0	0	0	3	6	9	12	15	18	21	21	21	-	-
9	0	0	0	0	0	0	0	0	1	2	3	4	5	6	7	8	-	-
10	0	0	0	0	0	0	0	0	0	2	4	6	6	6	6	6	-	-
11	0	0	0	0	0	0	0	20	40	60	60	60	60	60	60	60	-	-
Total	49	77	105	133	161	189	333	380	408	434	440	446	450	454	455	456	456	456

AC	1	2	3	4	5	6	7	8	9	10	11	12	13	14	15	16	17	18
2	20	40	60	80	100	120	120	120	120	120	120	120	120	120	120	120	120	120
3	4	8	12	16	20	24	28	32	36	40	44	44	44	44	44	44	44	44
4	25	25	25	25	25	25	25	25	25	25	25	25	25	25	25	25	25	25
5	0	0	0	0	0	0	20	40	60	80	80	80	80	80	80	80	80	80
6	0	4	8	12	12	12	12	12	12	12	12	12	12	12	12	12	12	12
7	0	0	0	0	120	240	240	240	240	240	240	240	240	240	240	240	240	240
8	0	0	0	0	0	0	3	6	9	12	15	18	21	24	24	24	24	24
9	0	0	0	0	0	0	0	0	0	0	1	2	3	4	5	6	7	8
10	0	0	0	0	0	0	0	0	0	0	0	2	4	6	8	10	10	10
11	0	0	0	0	0	0	20	40	60	80	80	80	80	80	80	80	80	80
Total	49	77	105	133	277	421	468	515	562	609	617	623	629	635	638	641	642	643

EV	1	2	3	4	5	6	7	8	9	10	11	12	13	14	15	16	17	18
2	13.33	26.67	40	53.33	66.67	80	80	80	80	80	80	80	80	80	80	80	80	80
3	3.27	6.55	9.82	13.09	16.36	19.64	22.91	26.18	29.45	32.73	36	36	36	36	36	36	36	36
4	25	25	25	25	25	25	25	25	25	25	25	25	25	25	25	25	25	25
5	0	0	0	0	0	0	20	40	60	80	80	80	80	80	80	80	80	80
6	0	6.67	13.33	20	20	20	20	20	20	20	20	20	20	20	20	20	20	20
7	0	0	0	0	60	120	120	120	120	120	120	120	120	120	120	120	120	120
8	0	0	0	0	0	0	2.62	5.25	7.88	10.5	13.12	15.75	18.38	21	21	21	21	21
9	0	0	0	0	0	0	0	0	0	0	1	2	3	4	5	6	7	8
10	0	0	0	0	0	0	0	0	0	0	0	1.2	2.4	3.6	4.8	6	6	6
11	0	0	0	0	0	0	15	30	45	60	60	60	60	60	60	60	60	60
Total	41.6	64.89	88.15	111.42	188.03	264.64	305.53	346.43	387.33	428.23	435.12	439.95	444.78	449.6	451.8	454	455	456

	1	2	3	4	5	6	7	8	9	10	11	12	13	14	15	16	17	18
SV	-7.39	-12.12	-16.85	-21.58	27.03	75.64	-27.47	-33.57	-20.67	-5.77	-4.88	-6.05	-5.23	-4.4	-3.2	-2	-1	0
CV	-7.39	-12.12	-16.85	-21.58	-88.97	-156.36	-162.47	-168.57	-174.67	-180.77	-181.88	-183.05	-184.23	-185.4	-186.2	-187	-187	-187
SPI	0.85	0.84	0.84	0.84	1.17	1.4	0.92	0.91	0.95	0.99	0.99	0.99	0.99	0.99	0.99	1	1	1
CPI	0.85	0.84	0.84	0.84	0.68	0.63	0.65	0.67	0.69	0.7	0.71	0.71	0.71	0.71	0.71	0.71	0.71	0.71
ES	0.85	1.57	2.4	3.23	5.97	6.53	6.81	7.29	8.26	9.78	10.19	10.99	11.8	12.9	13.45	14	15	16
SV(t)	-0.15	-0.43	-0.6	-0.77	0.97	0.53	-0.19	-0.71	-0.74	-0.22	-0.81	-1.01	-1.2	-1.1	-1.55	-2	-2	-2
SPI(t)	0.85	0.78	0.8	0.81	1.19	1.09	0.97	0.91	0.92	0.98	0.93	0.92	0.91	0.92	0.9	0.88	0.88	0.89

results on higher WBS levels, which greatly simplify final duration and completion date forecasting.

The future chapters will build further on the issues presented in this chapter. The three duration forecasting methods are discussed in detail and tested on a small set of real life project instances in chapter 3. In order to generalize the results found in the study, the methods will also be tested on a large and diverse set of fictitious projects based on a full factorial simulation experiment in chapter 4. Chapter 5 aims at linking the accuracy of the forecasting methods to their corresponding corrective actions that might be necessary in case of project problems, and investigates the ability of activity sensitivity information to guide and improve the corrective actions decision making process. Finally, chapter 6 compares the efficiency of two alternative project tracking methods, using the results obtained in the two previous chapters. All test results presented in the next chapters have been programmed in a novel software tool, ProTrack, which is the topic of chapter 7.

As a final remark, the letter to the editor of Harvard Business Review from Cooper (2003) as a response to the article written by Fleming and Koppelman (2003) is cited. In this letter, the author argues that the use of earned value management can be questioned when they are applied in highly complex projects. Due to the cycles of rework, the accuracy of the EVM metrics can be biased, leading to incorrect management decisions. In this book, it is not the ultimate goal to investigate this research topic into detail and/or to provide a complete answer on this issue. However, the next chapter will provide a partial answer on this issue as initiated by Lipke (2004) who measures the effective earned value when the project is the subject to a vast amount of rework cycles.

Chapter 2
Beyond the EVM Fundamentals

Despite the ever growing positive attention to EVM, the use of earned value metrics has been questioned by various authors when it is applied to highly complex and/or innovative projects. Indeed, EVM rests on a fundamental assumption that project activities and precedence relations are known in advance, and hence, assumes a project setting where estimates (activity duration, resource requirements, unexpected events, etc.) can be given within a certain range. However, projects often do not fulfil these assumptions but, on the contrary, are commonly plagued by fundamentally unforeseeable events and/or unknown interactions among various actions and project parts (Loch et al, 2006). The straight application of the standard EVM metrics and methods is certainly insufficient for projects where these assumptions do not hold. We refer to the book by Loch et al (2006) for a more general framework where the authors position and classify the sources of risk by the foreseeability of the underlying influence factors and by the complexity of the project. We also cite the letter to the editor of Harvard Business Review from Cooper (2003) as a response to the article written by Fleming and Koppelman (2003). In this letter, the author argues that the use of earned value management can be questioned when it is applied to highly complex projects. Due to the cycles of rework, the accuracy of the EVM metrics can be biased, leading to incorrect management decisions.

This chapter has no intention whatsoever to broaden or extend the EVM framework to *any* type of project behaving under *any* circumstance, but instead, investigates a recently developed extension of the earned schedule method (Lipke, 2003) and measures the influence of rework due to the lack of scheduling adherence on the forecasting performance of the existing EVM indicators. More precisely, the so-called p-factor concept, initially introduced by Lipke (2004) in the Measurable News, is discussed in detail in this chapter and will be critically reviewed in the simulation study of chapter 4.

M. Vanhoucke, *Measuring Time*, International Series in Operations Research & Management Science 136, DOI: 10.1007/978-1-4419-1014-1_2,

2.1 The p-factor concept for schedule adherence

Lipke (2004) has proposed a new earned value based measure to provide the connection of project output to EVM. He has defined the p-factor as the portion of earned value accrued in congruence with the baseline schedule, i.e. the tasks which ought to be either completed or in progress. The rationale behind this new measure is that performing work not according to the baseline schedule often indicates activity impediments or is likely a cause of rework. The basic assumption behind this new approach lies in the idea that, whenever impediments occur (activities that are performed relatively less efficiently compared to the project progress), resources are shifted from these constrained activities to other activities where they could gain earned value. However, this results in a project execution which deviates from the original baseline schedule. Consequently, this might involve a certain degree of risk, since the latter activities are performed without the necessary inputs, and might result into a certain portion of rework. The p-factor measures this schedule adherence and relies on the earned schedule concept calculated as follows:

$$p = \frac{\sum_{i \in N} min(\mathrm{PV}_{i,\mathrm{ES}}, \mathrm{EV}_{i,\mathrm{AT}})}{\sum_{i \in N} \mathrm{PV}_{i,\mathrm{ES}}} \tag{2.1}$$

N	Set of activities in a project
$\mathrm{PV}_{i,\mathrm{ES}}$	Planned Value of activity *i* at time instance ES
$\mathrm{EV}_{i,\mathrm{AT}}$	Earned Value of activity *i* at the current time AT

The p-factor is the ratio of the earned value corresponding to the baseline schedule divided by the total planned value at time instance ES. Since the nominator takes the minimum of the planned value at time unit ES and the earned value accrued at the actual time, the p-factor obviously always lies between zero and one, inclusive. Hence, the p-factor measures to what degree the earned value is accrued according to the baseline schedule (100% means a perfect schedule adherence).

Figure 2.1 displays a fictitious real life performance of the example project network of figure 1.7, resulting in a real duration RD of 17 time units (a project delay of one time unit). The figure displays the current state of the project at the actual time of 7, and shows that some activities are not performed in congruence with the baseline schedule. More precisely, there is a delay for activity 3 from 9 to 12 time units (currently, the percentage completed at AT = 7 equals $\frac{7}{12}$ = 58,33%) and for activity 7 from 1 to 4 time units (percentage completed equals $\frac{1}{4}$ = 25%). Furthermore, there are overlaps between activities 2 and 5, 5 and 9 and 7 and 11 with 2, 1 and 1 time units, respectively. The earned schedule value equals ES = 6.48 denoting a project delay, i.e. SV(t) = 0.52 time units. Details on the calculation of the ES value can be found in the upper part of table 2.2 displayed at the end of this chapter.

As an example, the baseline schedule of figure 1.8 shows that PV_6 = 100% * 80 + 66.66% * 36 + 100% * 25 + 50% * 80 + 100% * 20 = 189 and PV_7 = 100% * 80 + 77.77% * 36 + 100% * 25 + 75% * 80 + 100% * 20 + 100% * 120 = 333. The black bars of figure 2.1 show that EV_7 = 100% * 80 + 58.33% * 36 + 100% * 25 +

100% * 80 + 100% * 20 + 25% * 120 + 25% * 8 = 258. The earned schedule metric ES identifies the time at which the amount of earned value EV accrued should have been earned and consequently, since $PV_6 < EV_7 < PV_7$, $ES_7 = 6 + \frac{258-189}{333-189} = 6.48$. The ES metric at the current time AT = 7 is shown in figure 2.2.

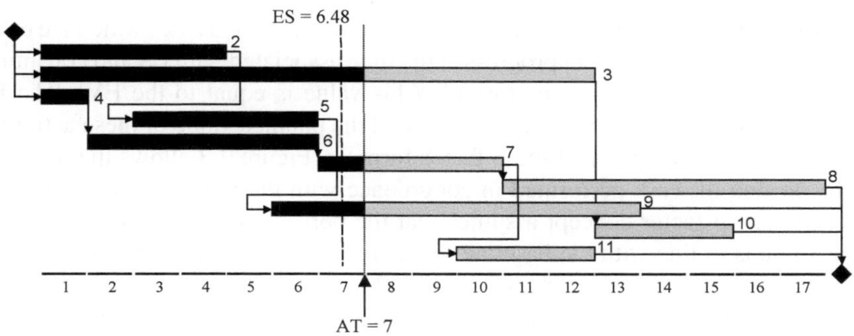

Fig. 2.1 Real life execution of the example project (RD = 17)

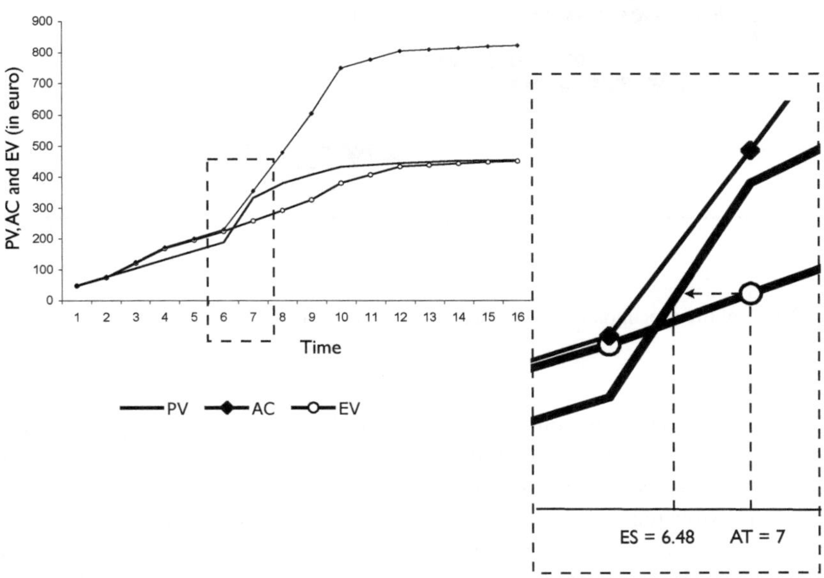

Fig. 2.2 The ES metric at current time AT = 7

Figure 2.3 displays the EV accrue of the example project at time AT = 7 relative to the baseline schedule. In this format, the progress of the project activities is shown relative to the baseline schedule by filling the planned durations of the baseline schedule according to the percentage completed. Graphically, the p-factor is equal to the earned value (black bars) to the left of the earned schedule line divided by the total EV. The ES line is defined as the point on the baseline schedule where PV is equal to EV and consequently the ES value is the time at which the EV should have been accrued if the baseline schedule had been followed. The value of PV at time instance ES is the sum of planned values from a set of activities which are either planned to be completed or in progress at the time associated with ES and constitutes the denominator of the p-factor formula (this value is equal to the EV). The EV corresponding to these activities, not to exceed the planned value of these activities, is the EV in the numerator of the p-factor formula. Figure 2.3 allows the detection of the portion of work performed in congruence with the baseline schedule. More precisely, the p-factor concept assumes that the portion of work performed before the ES line is in line with the baseline schedule while the portion of work after the ES line is performed under risk (see section 2.2). Consequently, the p-factor at AT = 7 is equal to $\frac{225.58}{257.99} = 0.87$, denoting that the project performance is not 100% in line with the baseline schedule of figure 1.8. The calculations can be found in table 2.1.

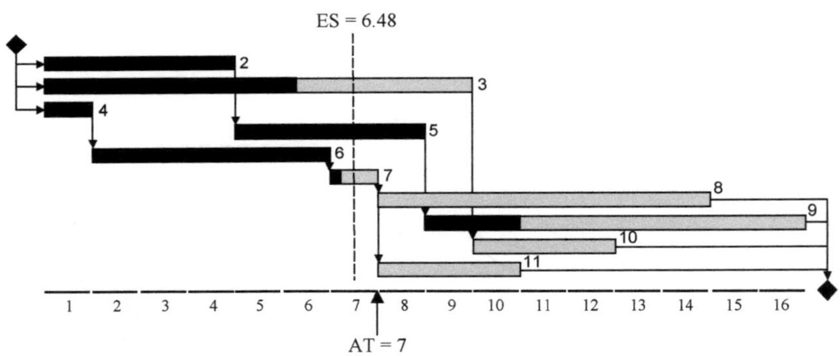

Fig. 2.3 Real life execution of the example project relative to the baseline schedule

Section 2.2 elaborates on the use of the p-factor to reveal the project impediments or constraints causing inefficient use of resources and to detect the portion of work performed that is sensitive to rework. In the remainder of this section, three types of possible reasons that cause lack of schedule adherence are briefly discussed, i.e. activity overlapping, difference between the PV and EV accrue and unexpected changes in the original activity time estimates. These causes will be implemented in the second simulation model in section 4.3 of chapter 4.

Table 2.1 The p-factor calculation for the example project execution of figure 2.3

i	PC	$PV_{i,6.48}$	$EV_{i,7}$	$\min(PV_{i,6.48};EV_{i,7})$
2	100%	80	80	80
3	58.33%	25.92	21	21
4	100%	25	25	25
5	100%	49.58	80	49.58
6	100%	20	20	20
7	25%	57.49	30	30
9	25%	0	2	0
Sum		257.99		225.58

2.1.1 Activity overlapping

Activities are often performed out-of-sequence which results in activities that start before the finish of at least one of its predecessors (further denoted as activity overlapping, but this is also known as concurrent engineering, integrated product development or fast tracking). Consequently, these project activities are executed without the complete information input they need from their predecessor activities which engenders a certain level of risk. This increased risk level implies the presence of iterations, loops and/or rework, and hence leads to loss in time and/or cost. Measuring rework due to activity overlapping is not new and has been studied before in literature. Roemer et al (2000) investigate the trade-off between product development time and costs and determine the optimal overlapping strategy under different scenarios. They consider the expected design time and cost increase due to overlapping as a function of the overlap duration between two design stages. They calculate the probability of rework as a function of the information exchange between two consecutive stages. More precisely, they measure the so-called evolution as the speed of reliable information exchange between two stages and the sensitivity as the maximal probability of rework due to overlapping. In the example execution, there is activity overlapping between activities 2 and 5, 5 and 9 and 7 and 11 with 2, 1 and 1 time units, respectively (see figure 2.1).

2.1.2 EV/PV accrue deviation

Activities that are completed within their estimated time and budget are not necessarily performed in congruence with their predefined planned value. The choice of a PV accrue needs to be done before the start of the project, and can be chosen from various earned value measurement methods (e.g. the percent complete estimates, the milestones with weighted values, fixed formulas such as 25/75, 50/50 and many more. For an overview, the reader is referred to a standard EVM handbook or to the summary table 7.1 of chapter 7). In this book, the earned value measurement method is assumed to follow a linear accrue which ignores (positive and/or negative) learning effects, unless otherwise indicated. In a real project environment it is seldom true that the planned values are linear. Instead, one can assume a learning curve factor to denote that work efficiency increases or decreases over time due to experience and other beneficial factors (positive learning) or a higher degree of complexity (negative learning). Learning curves have been studied in literature from a project scheduling and monitoring point of view by Amor (2002), Amor and Teplitz (1993), Amor and Teplitz (1998), Badiru (1995), Shtub (1991) and Shtub et al (1996). As a result, the earned value accrue often deviates from the planned value accrue, even when the activity real terminal (time and cost) values are equal to the planned values, leading to a lack of 100% schedule adherence.

2.1.3 Ahead or delays in activities

Obviously, inaccuracies in the initial activity time and cost estimates lead to deviations from the baseline schedule. In chapter 4, variation in activity durations will be simulated under various settings. In the example execution of figure 2.1, activities 3 and 7 show a delay of 3 time units compared to its baseline duration.

2.2 Rework due to lack of schedule adherence

The main contribution of the p-factor is to serve as a management tool to reveal impediments or constraints during the execution of the project and to detect the portion of work performed under risk. The fundamental idea behind the p-factor concept lies in the detection of impediments or constraints whenever resources are used in an inefficient way resulting in activity execution that lags the ES performance. This inefficiency, expressed by the p-factor, allows to correct the accrued earned value to the risk of rework, and calculates the *effective earned value* EV(e) as the risk-adapted portion of earned value that is performed within the expected baseline schedule performance, defined by the earned schedule metric.

Figure 2.4 illustrates the concept of effective earned value EV(e) on the example project at time AT = 7. The ES value splits the accrued earned value into two disjoint subsets (the black bars to the left and to the right of the ES line) which both need a different interpretation. On the one hand, the portion of the work (i.e. the accrued earned value) to the left of the ES line is assumed to be performed without risk. However, the figure shows that activities 3 and 7 are performed less efficiently relative to the normal project progress as measured by the ES line, which might indicate the presence of an impediment or project constraint. The portion of work, on the other hand, to the right of the ES line indicates work which is ahead of the normal project performance and is assumed to have a certain degree of risk. This degree of risk is often the result of inefficient use of resources which were shifted from the constrained activities (in this case 3 and 7) to less constrained activities where the resources could gain earned value. However, since these resources often work without the necessary inputs, their work might result into a certain portion of rework (i.e. risk). In order to take this risk of rework into account, a correction to the earned value is needed to distinguish between the portion without risk (EV(p)) and the portion of work which is likely performed under risk (EV(r)).

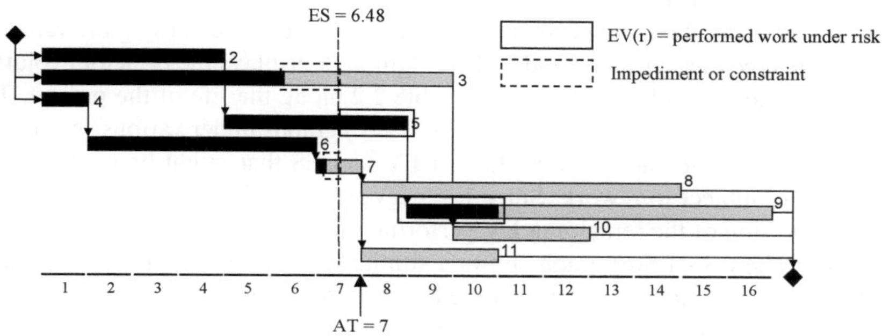

Fig. 2.4 The p-factor reveals activity impediments and work performed under risk

Consequently, the p-factor splits the total earned value at the current actual time into a risk-free fraction of work performed in congruence with the baseline schedule (EV(p)) and a portion of work performed under risk (EV(r)), i.e.

$$\mathrm{EV} = \mathrm{p} * \mathrm{EV} + (1 - \mathrm{p}) * \mathrm{EV} = \mathrm{EV(p)} + \mathrm{EV(r)} \tag{2.2}$$

with

EV	Earned value
EV(p)	Risk-free earned value
EV(r)	Remaining earned value portion performed under risk

Since the portion of earned value under risk is likely to cause rework, the effective earned value EV(e) decreases the total earned value to take the possibility of rework into account, as follows:

$$\mathrm{EV(e)} = \mathrm{EV(p)} + \mathrm{R\%} * \mathrm{EV(r)} \tag{2.3}$$

with

EV(e)	Effective earned value
R%	Estimated portion of EV(r) that is usable and requires no rework

EV(e) can be simplified to

$$\mathrm{EV(e)} = \mathrm{p} * \mathrm{EV} + \mathrm{R\%}(1 - \mathrm{p})\mathrm{EV} = \mathrm{EV}(\mathrm{p} + \mathrm{R\%} - \mathrm{p} * \mathrm{R\%}) \tag{2.4}$$

and consequently, the plan adherence (i.e. the p-factor), along with rework, and cumulative earned value determine the effective earned value (Lipke, 2004).

Note that, since EV(e) $\leq$ EV, the following equations hold:

1. SPI(e) $\leq$ SPI
2. ES(e) $\leq$ ES which means that SPI(t)(e) $\leq$ SPI(t)

Consequently, the p-factor and the resulting effective earned value can be considered as a correction factor to account for possible rework by decreasing the values of the performance measures SPI and SPI(t). Figure 2.5 displays the p-factor evolution of the example network as calculated in table 2.2 along the life of the project. The figure also displays the EV - EV(e) and ES - ES(e) graphs under various settings for the R% measure and shows the portion of EV and ES that might be unusable and might be the subject to rework. Since EV - EV(e) = (1 - R%) * EV(r), the graphs show the portion of the remaining EV performed under risk that will be the subject to rework. The ES - ES(e) graph shows a similar portion of rework, but translated into time units. Obviously, the higher the R%, the lower the portion of EV(r) that is unusable and hence, the lower the risk of rework.

Note that this notion of rework rests on a number of crucial assumptions and hence, is the subject to a number of limitations. In general, activities may require rework because they were done incorrectly or because the work or information they were based on was itself erroneous or has changed (Ford and Sterman, 2003). The current chapter, however, exploits a limited view on uncertainty and rework, and excludes rework due to quality problems (see e.g. Icmeli-Tukel and Rom (1997)), unknown activities in the project network (e.g. Pollack-Johnson and Liboratore (2005)) or uncertainty and chaos in general (Loch et al, 2006). Instead, it is restricted to the lack of adherence to the baseline schedule involving impediments and work under risk as a source of potential rework in part of the project activities. Moreover, unlike various sources in literature (see e.g. Icmeli-Tukel and Rom (1997)), the chapter assumes a discovery time for rework equal to zero, i.e. portions of activities that are ahead of the earned schedule progress are subject to potential risk which immediately results in activity delays. Consequently, it also ignores an extreme form of rework, the so-called 90% syndrome, as investigated by e.g. Ford and Sterman (2003). These authors claim that even projects that are staffed by skilled personnel with ample resources can experience the 90% syndrome, solely as a function of the informational and physical dependencies created by concurrency. The greater the overlap between activities, the more work is completed and released before rework requirements can be detected, leading to more unplanned iterations. Next, the current view on activity overlapping assumes a unilateral information exchange without loops, branches or iterations as in the GERT approach (Moder et al, 1983) and without the presence of a bi-directional information flow as is the case in the Design Structure Matrix (DSM) approach (Steward, 1981). The p-factor concept of this chapter also makes abstraction from the increasing body of literature dedicated to overlapping in product design, which is basically divided into two overlapping views (Roemer and Ahmadi, 2004), i.e. overlapping as a necessary and promising decision to avoid costly late changes or product obsolescence and overlapping as a calculated

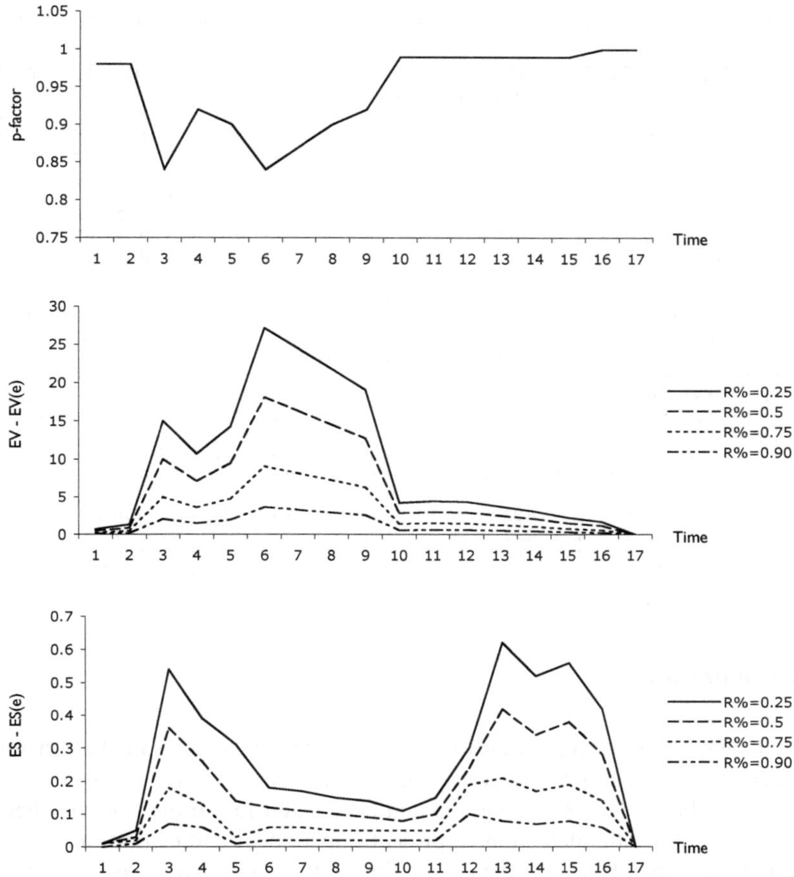

Fig. 2.5 The evolution of the p-factor and the EV - EV(r) and ES - ES(e) evolution for various R% values

risk to reduce the project duration. In the current manuscript, activity overlapping is allowed to reduce project duration, and involves a project performance that deviates from the baseline schedule possibly leading to rework. To that respect, the current study follows the second view on activity overlapping, and considers it as a well-considered form of risk taken to reduce project duration. The reader is referred to a brief overview presented by Roemer et al (2000) who investigate the trade-off between increased cost and reduced project duration due to overlapping. An extensive overview and a further discussion on product design overlapping is outside the scope of this book.

Table 2.2 The periodically earned value based measures for the example project (R% = 0.9)

	1	2	3	4	5	6	7	8	9	10	11	12	13	14	15	16	17
	Traditional Earned Value Calculations																
EV	48	75	122	169	196	224	258	292	326	380	407	434	440	445	450	453	456
AC	49	77	125	173	201	230	355	480	605	750	778	806	812	817	822	825	828
PV	49	77	105	133	161	189	333	380	408	434	440	446	450	454	455	456	456
SV	-1	-2	17	36	35	35	-75	-88	-82	-54	-33	-12	-10	-9	-5	-3	0
CV	-1	-2	-3	-4	-5	-6	-97	-188	-279	-370	-371	-372	-372	-372	-372	-372	-372
SPI	0.98	0.97	1.16	1.27	1.22	1.19	0.77	0.77	0.8	0.88	0.93	0.97	0.98	0.98	0.99	0.99	1
CPI	0.98	0.97	0.98	0.98	0.98	0.97	0.73	0.61	0.54	0.51	0.52	0.54	0.54	0.54	0.55	0.55	0.55
ES	0.98	1.93	3.61	5.29	6.05	6.24	6.48	6.72	6.95	8	8.96	10	11	11.83	13	13.75	16
SV(t)	-0.02	-0.07	0.61	1.29	1.05	0.24	-0.52	-1.28	-2.05	-2	-2.04	-2	-2	-2.17	-2	-2.25	-1
SPI(t)	0.98	0.96	1.2	1.32	1.21	1.04	0.93	0.84	0.77	0.8	0.81	0.83	0.85	0.85	0.87	0.86	0.94
	Effective Earned Value Calculations																
p-factor	0.98	0.98	0.84	0.92	0.9	0.84	0.87	0.9	0.92	0.99	0.99	0.99	0.99	0.99	0.99	1	1
EV(e)	47.91	74.83	120	167.6	194.1	220.4	254.8	289.1	323.5	379.5	406.4	433.5	439.5	444.6	449.7	452.8	456
SV(e)	-1.09	-2.17	15	34.57	33.1	31.39	-78.2	-90.9	-84.5	-54.5	-33.6	-12.5	-10.5	-9.42	-5.3	-3.22	0
CV(e)	-1.09	-2.17	-5	-5.43	-6.9	-9.61	-100	-191	-282	-371	-372	-373	-373	-372	-372	-372	-372
SPI(e)	0.98	0.97	1.14	1.26	1.21	1.17	0.77	0.76	0.79	0.87	0.92	0.97	0.98	0.98	0.99	0.99	1
CPI(e)	0.98	0.97	0.96	0.97	0.97	0.96	0.72	0.6	0.53	0.51	0.52	0.54	0.54	0.54	0.55	0.55	0.55
ES(e)	0.98	1.92	3.54	5.23	6.04	6.22	6.46	6.7	6.93	7.99	8.94	9.98	10.92	11.76	12.92	13.69	16
SV(t)(e)	-0.02	-0.08	0.54	1.23	1.04	0.22	-0.54	-1.3	-2.07	-2.01	-2.06	-2.02	-2.08	-2.24	-2.08	-2.31	-1
SPI(t)(e)	0.98	0.96	1.18	1.31	1.21	1.04	0.92	0.84	0.77	0.8	0.81	0.83	0.84	0.84	0.86	0.86	0.94

2.3 Conclusion

In this chapter, a new concept in earned value management to measure the adherence of project performance to the baseline schedule has been reviewed. This novel concept, the so-called p-factor, is a simple and easy extension to the earned schedule approach and allows the detection of project impediment and/or portions of work performed under risk based on the calculation of the traditional earned value metrics.

This new approach is another step in simplifying the often time-consuming and burdensome project tracking effort of a project manager during project progress, and closely connects to the remarks made in chapter 1 when discussing the criticism and remarks of Jacob and Kane (2004) on the ideal WBS level of control when using EVM. Indeed, although some EVM practitioners still hold the belief that schedule analysis can only be correctly accomplished through detailed project network and critical path analysis (i.e. at the lowest level of the WBS), EVM is mainly set up as an easy and quick sanity check at higher WBS levels to measure, by means of straightforward schedule performance indicators, the current project time performance and to forecast the final project duration.

The main intention of chapter 4 is to test the validity and reliability of this higher WBS level approach of the schedule performance indicators (SPI and SPI(t)) and the potentially powerful p-factor indicator used in EVM. Unlike the earned schedule method, presented in the previous chapter, the p-factor method and its potential to measure schedule adherence have not been tested widely on real life data. Therefore, future research is encouraged. To that purpose, section 4.3 of chapter 4 presents a detailed simulation study in which the network structure, the degree of activity overlapping, the earned value accrue and the variation in the activity duration have been carefully controlled. The simulation model presented there aims at measuring the forecast accuracy of various duration forecasting measures with and without the use of the p-factor.

Chapter 3
A Case Study

In a world where an unattended suitcase is a threat, baggage handling is a matter not just of convenience, but also of security. Fabricom Airport Systems provides the technology that sorts and screens passenger luggage safely. The company designs and develops online 100% hold baggage screening (HBS) systems and supplies baggage handling systems to regional and international airports around the globe. Over the last 40 years, Fabricom Airport Systems and its subsidiaries have been responsible for designing and implementing over 400 baggage handling solutions, including 80 HBS systems[1]. In line with corporate restructuring, Fabricom Airport Systems has taken on the new identity of Logan Teleflex. This move recalls brands that have been synonymous throughout the world for high performance and quality baggage handling systems for many years.

The illustrations and results of this section are drawn from an implemented earned value management approach for managing complex system projects of an airport luggage handling system at Fabricom Airport Systems in Brussels (Belgium). At the time of the study, the projects were carried out under Fabricom Airport Systems, and hence, this name will be used throughout the chapter. Figure 3.1 shows an illustration of the luggage handling system at the Belgian airport at Brussels. Weekly meetings with the project team provided the progress data, which were then translated into earned value metrics, according to the predefined earned value methods. The data were then rolled up to monthly values for formal project performance reporting. All calculations and graphs have been made in Microsoft Excel spreadsheets. The different schedule forecasting methods will be applied to real project data for three projects. Each project has a different performance: one project is behind schedule but under cost, one project is late with a cost overrun and one project is ahead of schedule but with a cost overrun. The real life data of the three projects are summarized in table 3.1.

Figure 3.2 displays the dashboard used for the time/cost performance evaluation of the three projects over time. The dashboard displays the schedule performance (SPI or SPI(t)) on the x-axis and the cost performance (CPI) on the y-axis. In the

[1] Source: "Fabricom Airport Systems: A case for security", The Manufacturer, May 2005.

M. Vanhoucke, *Measuring Time*, International Series in Operations Research & Management Science 136, DOI: 10.1007/978-1-4419-1014-1_3,

Fig. 3.1 The luggage handling system at Brussels Airport, Zaventem, Belgium

Table 3.1 Real life project data for 3 projects at Fabricom Airport Systems

	Project	Category	Budget at Completion	Cost at Completion	Planned Duration	Actual Duration
1	Revamp Check-in	Late Finish Cost Underrun	€ 360,738	€ 349,379	9	13
2	Link Lines	Late Finish Cost Overrun	€ 2,875,000	€ 3,247,000	9	12
3	Transfer Platform	Early Finish Cost Overrun	€ 906,000	€ 932,000	10	9

remainder of this chapter, the three real life projects are briefly discussed and the EVM data are given in illustrative figures and tables.

Figure 3.3 shows the evolution of the three projects from their start to finish on the EVM schedule/cost dashboard measured by the SPI(t) and CPI. Figure 3.4 shows the evolution of the three projects from their start to finish on the EVM schedule/cost dashboard measured by the SPI and CPI. The grey dots show the final project time performance of the three projects copied from figure 3.3, which is an indication of the real time/cost performance at the end of the project. The figure clearly shows that the final cost performance (CPI) is similar for the two dashboards, but the time performance of the SPI indicator is completely false and gives an indication that all projects finish on time, while the real final time performance is given by the

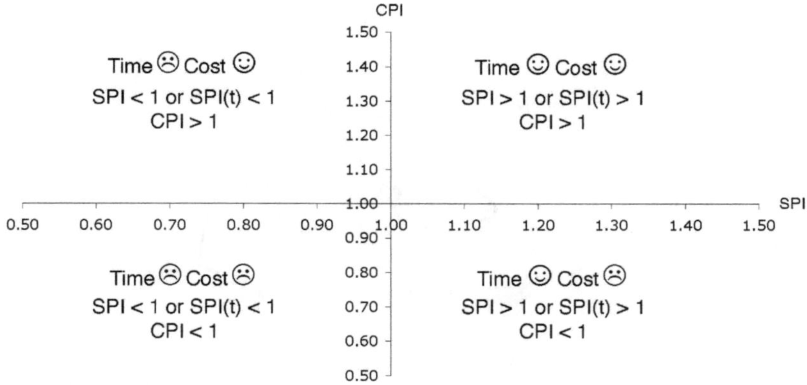

Fig. 3.2 The EVM schedule/cost performance dashboard

grey dots. More details and graphs on the individual projects will be given in the remainder of this chapter.

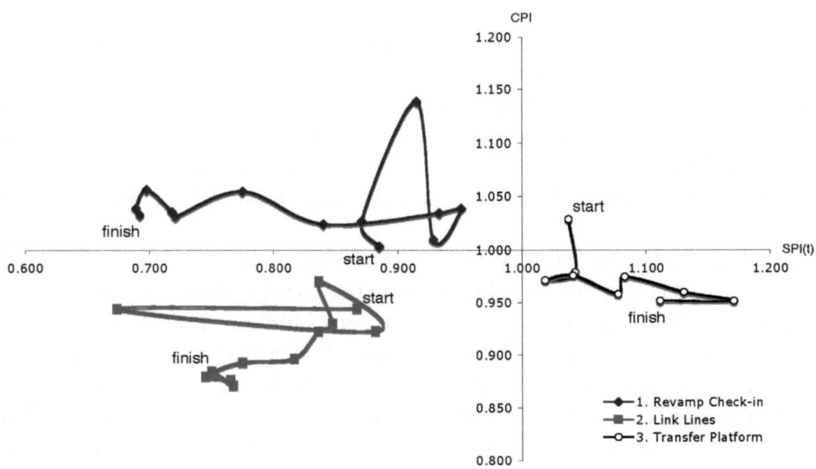

Fig. 3.3 The SPI(t)/CPI dashboard for the three example projects

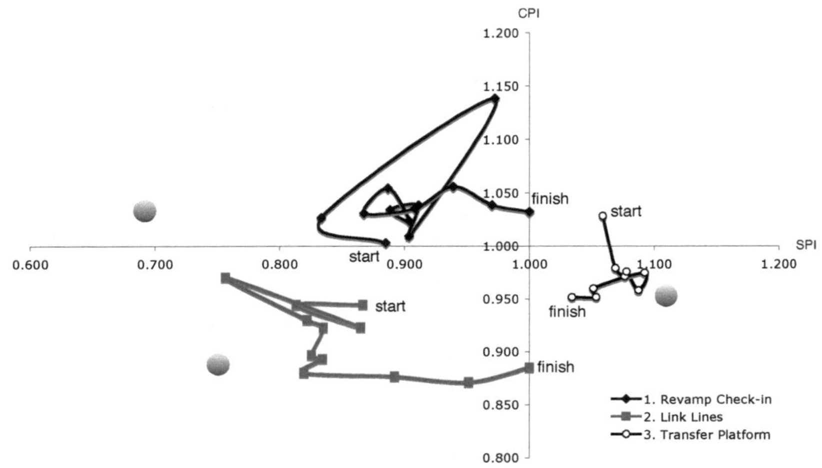

Fig. 3.4 The SPI/CPI dashboard for the three example projects

3.1 Project 1. Revamp check-in

The project concerns a revamping of different check-in islands. This project existed mainly out of electrical works (engineering, installation and commissioning) and automation works (programming, implementing and commissioning). The planned duration was 9 months, with a budget at completion of € 360,738. Detailed project data can be found in table 3.2. Figure 3.5 displays the schedule variances and schedule performance indices. The project was delivered 4 months later than expected, but under budget.

The graph of the SV and SV(t) along the project duration (the upper graph of figure 3.5) reveals that the SV follows a negative trend till February 2003, followed by a positive trend and finally ending with a zero variation. The SV(t) graph, on the contrary, shows a negative trend along the complete project duration, and ends with a cumulative variation of -4 months, which is exactly the projects delay. A similar effect is revealed in the graph of the schedule performance metrics (the bottom graph of figure 3.5). During the early and middle stages, both SPI and SPI(t) correlate very well. However, towards the late project stage (at the $\pm 75\%$ completion point), the SPI becomes unreliable showing an improving trend while the project is slipping further away. This further performance decline is clearly shown by the SPI(t) indicator.

The forecast of the three different schedule forecasting methods have been applied and displayed in figure 3.6. Each graph displays the three methods, with a different performance factor PF (see table 1.2) as follows: the top graph has a PF = 1, the middle graph has a PF = SPI or SPI(t) and the bottom graph has a PF = SCI or SCI(t). The graph reveals some repetitive patterns, regardless of the perfor-

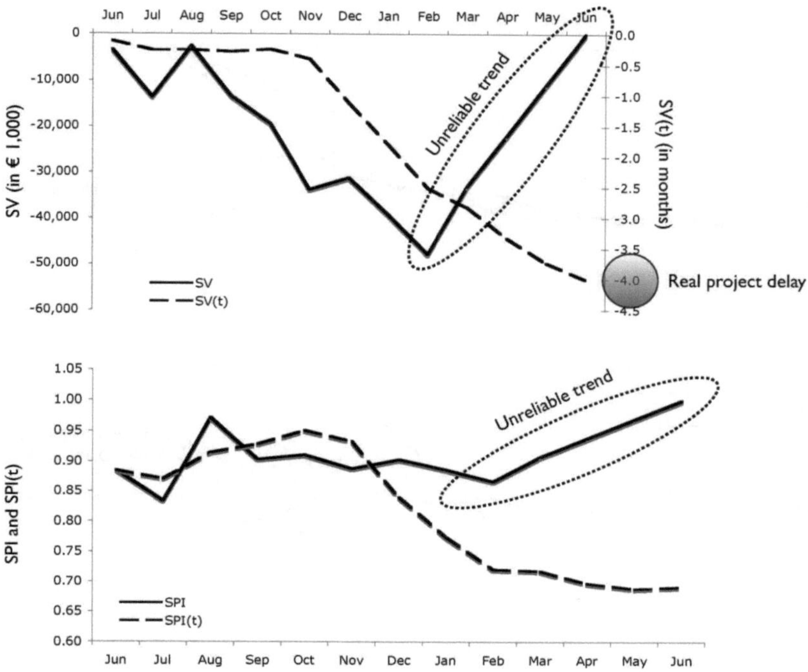

Fig. 3.5 Schedule performance measures for project 1 "Revamp check-in" (late finish, under budget) at Fabricom Airport Systems

mance factor PF. First, all methods correlate very well during the early and middle project stages, and produce nearly similar results. Second, the earned schedule method clearly outperforms all other methods during the last project stage reporting periods. Finally, the graphs display bizarre and unreliable results for the planned value method once the planned time at completion has been reached, and is therefore not a good time predictor for this project. The graphs also reveal that the earned schedule method almost always forecasts a higher project duration compared to the two other methods, for the three performance factors. Moreover, all methods are quasi insensitive to the PF value, which might be explained by the fact that the bad schedule performance (late finish) is compensated by a good cost performance (cost underrun).

The graph in figure 3.7 shows the evolution of the "to complete schedule performance indices" (which are only defined in the earned schedule and the earned duration method) over time. These indices show the performance needed to complete the project on time and is given by TCSPI (calculated with the earned duration method) and TCSPI(t) (calculated with the earned schedule method). At the early project stage, both indices produce similar results. Since there were no signs of an

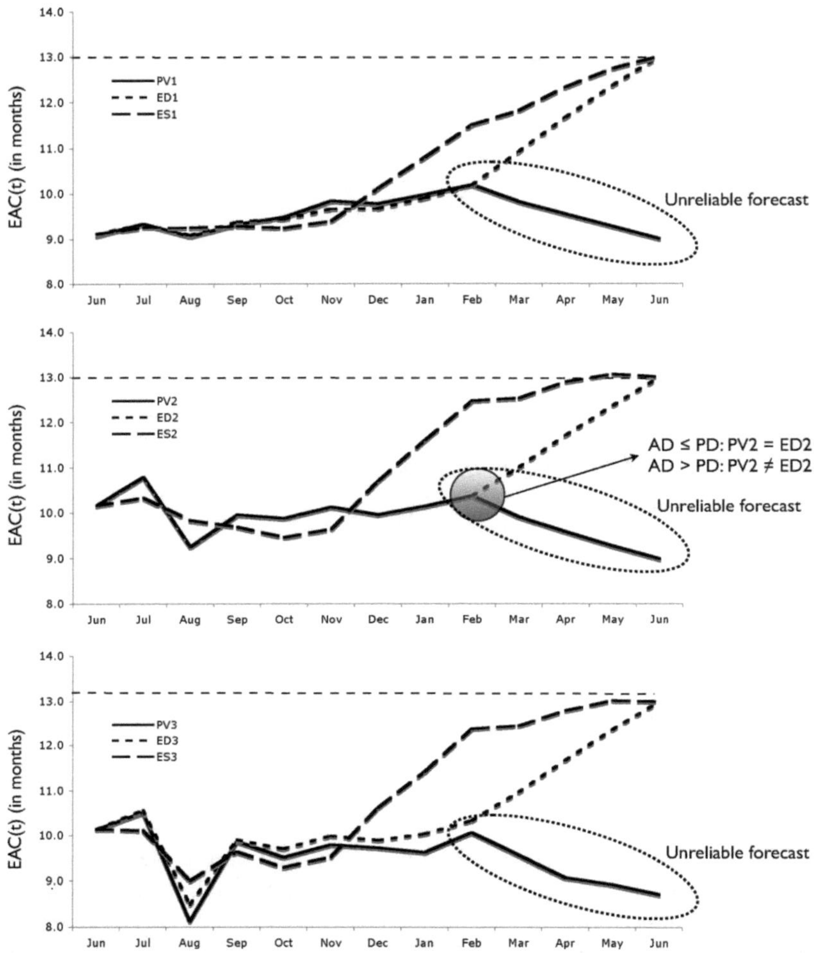

Fig. 3.6 Duration forecasting measures for project 1 "Revamp check-in" (late finish, under budget) at Fabricom Airport Systems

improved schedule efficiency at the September 2002 project review, it was decided to take a two-months project delay into account (revised project duration = 9 + 2 = 11 months). From this point onwards, the new TCSPI indicators (referred to as TCSPI - 11 and TCSPI(t) - 11 which correspond to the $TCSPI_{LRS}$ and $TCSPI(t)_{LRS}$ of chapter 1, with LRS = 11) have been computed. After 7 months (the December 2002 project review), the TCSPI - 11 indicator shows a declining trend, indicating that a lower performance efficiency is needed. However, the TCSPI(t) - 11 indicator just started an upward trend, which is a clear indication that improved performance

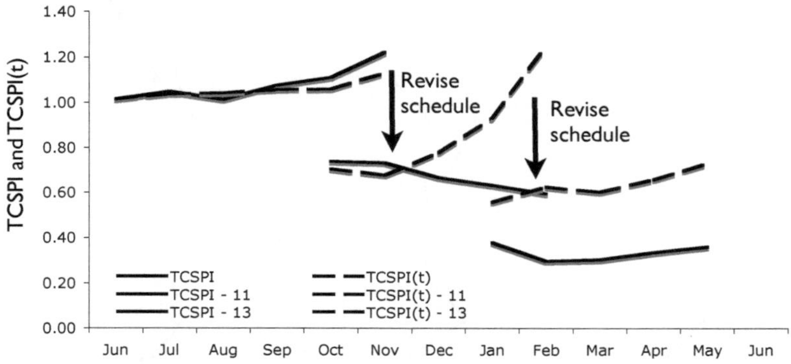

Fig. 3.7 The "to complete schedule performance index" for project 1 "Revamp check-in" (late finish, under budget) at Fabricom Airport Systems

is crucial to finish the project within the revised deadline of 11 months. A revised scenario to allow for a 4 months delay resulted in a revised targeted project duration of 13 months (with new indicators TCSPI - 13 and TCSPI(t) - 13). The TCSPI - 13 continuously shows a lower value compared to the TCSPI(t) - 13.

3.2 Project 2. Link lines

Table 3.3 displays the data of the link lines project, which links two piers with fully automated baggage conveying lines. The planned duration was 9 months, and the project finished 3 months later with a cost overrun. Figure 3.8 summarizes all EVM metrics in a similar way as figures 3.5, 3.6 and 3.7. The graphs reveal that the forecasting methods correlate well for the first two thirds of the project, and show a better performance of the earned schedule method towards the end of the project. A similar behavior is reflected in the "to complete schedule performance indices", producing similar results during the early project stages, and an outperforming accuracy for the TCSPI(t) - 11 index at the middle stage of the project. As a contrast, the TCSPI - 11 shows a stable trend (no increasing performance is needed) while the project keeps slipping further away.

3.3 Project 3. Transfer platform

The third project is a renovation of the transfer baggage conveying system due to changed baggage flows and security issues. This project had a planned duration

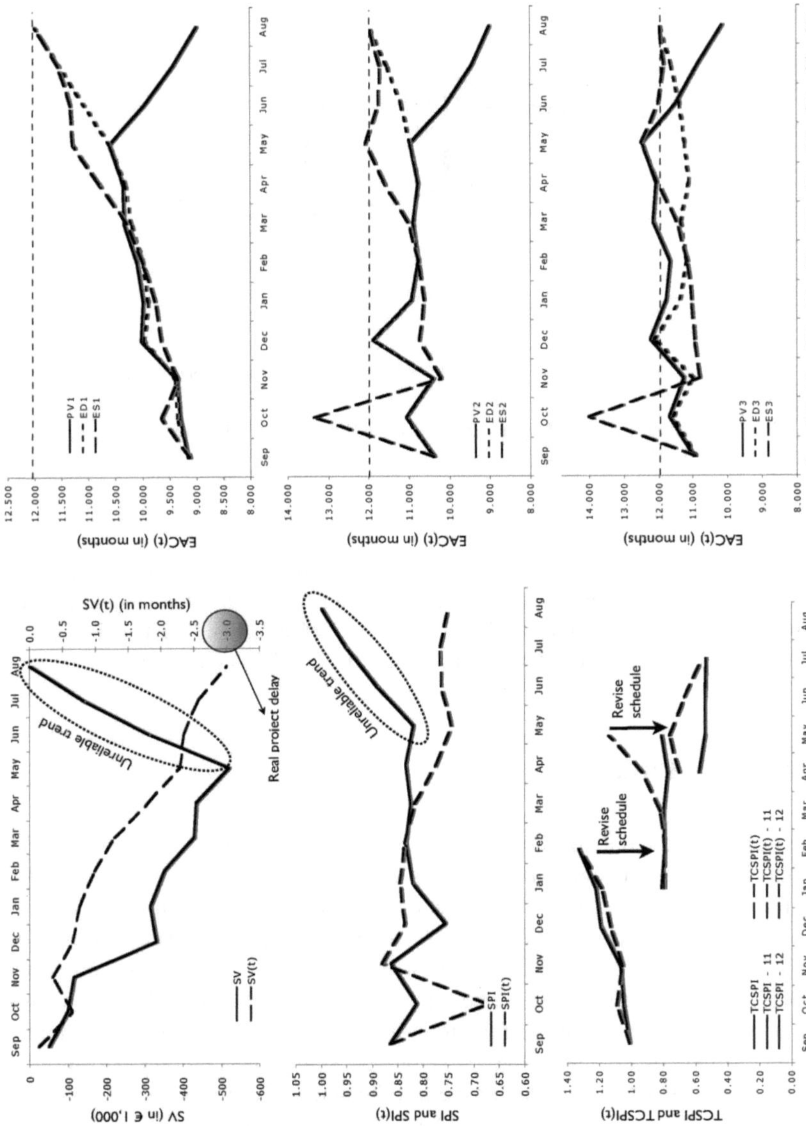

Fig. 3.8 The earned value metrics for project 2 "Link lines" (late finish, over budget)

of 10 months, while it finished within 9 months, with a cost overrun (see table 3.4 for details). The graphs of figure 3.9 confirm the results found by the previous projects. At the late project stage, the planned value and the earned duration methods give more pessimistic (i.e. longer duration) results and the TCSPI metrics produce higher values than the TCSPI(t). The overestimation of duration and/or the needed efficiency calculated by the planned value and earned duration methods may cause wrong decision making by the upper management.

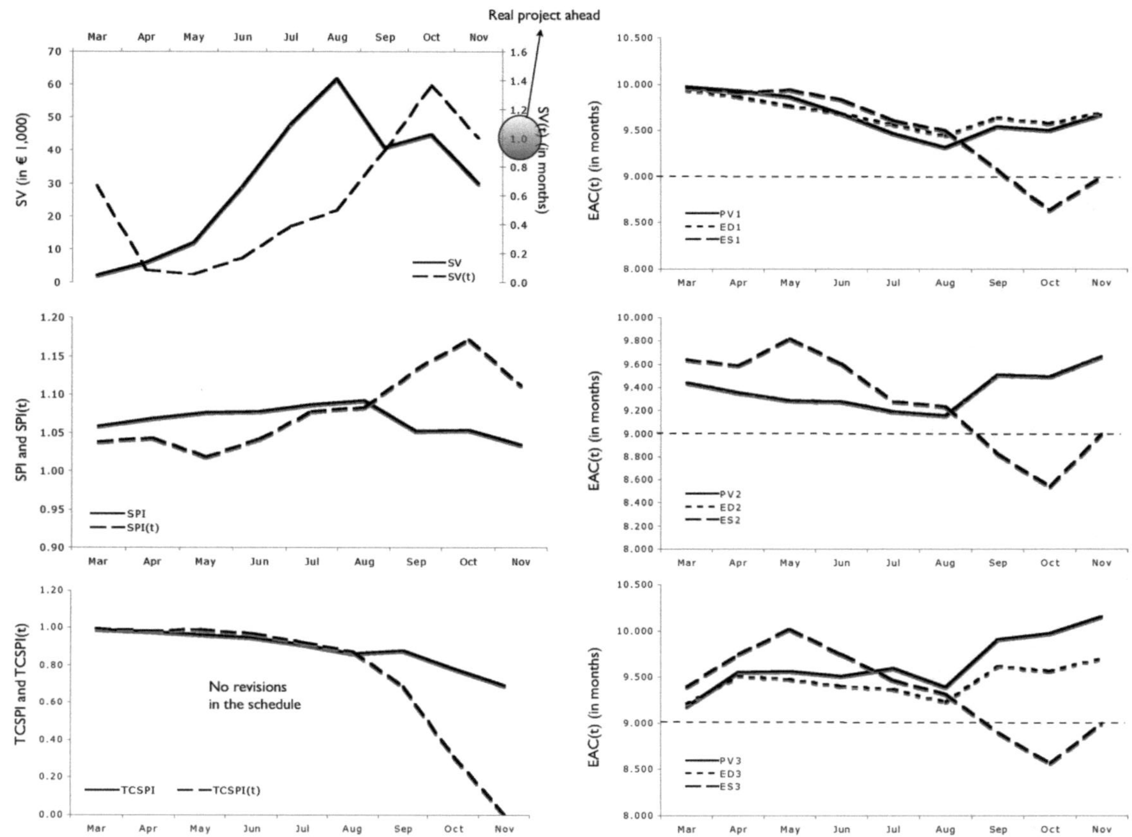

Fig. 3.9 The earned value metrics for project 3 "Transfer platform" (early finish, over budget)

3.4 Conclusion

In this chapter, the three different project duration methods presented in chapter 1 have been compared and evaluated on three real life projects from Fabricom Airport Systems, Belgium executed between July 2002 and August 2004. The results confirm the theoretical foundations and claims presented earlier and reveal that the earned schedule method was the only method which showed satisfying and reliable results during the whole project life cycle. Consequently, the case study confirms the claim that the results obtained by the planned value method and the earned duration method are unreliable at the end of the project. Instead, the earned schedule method seems to provide valid and reliable results along the project's lifespan.

Note that these results serve only as illustrations and no general conclusions can be drawn. Other similar empirical results have been drawn in various research papers. As an example, the validity of the earned schedule concept has been tested on a portfolio of six projects (Henderson, 2003) and on a small scale but time critical information technology software development project (Henderson, 2005). Hecht (2007) used data from a U.S. Navy project to build a helicopter trainer for maintenance personnel and used it as a case study to test the predictive power of the earned schedule method. Henderson and Zwikael (2008) give a summary of their project performance stability study by investigating a large set of projects from three different countries. Finally, Lipke et al (2008) have statistically validated the earned schedule concept on a large pool of real life projects.

Despite the necessity of empirical data to confirm theoretical claims and to validate novel methods, results are often very case-specific and depend on the (unknown) characteristics of the project. It is for this very reason that the simulation studies of the next chapters have been initiated. In doing so, more general results can be drawn for very different project networks under a wide variety of circumstances.

Table 3.2 Detailed information for project 1 "Revamp check-in" (cost in thousands of €)

	Jun-02	Jul-02	Aug-02	Sep-02	Oct-02	Nov-02	Dec-02	Jan-03	Feb-03	Mar-03	Apr-03	May-03	Jun-03
AC	25,567	66,293	78,293	124,073	191,367	259,845	285,612	290,843	303,489	316,431	320,690	336,756	349,379
EV	25,645	68,074	89,135	125,244	198,754	268,763	292,469	306,725	312,864	327,694	338,672	349,861	360,738
PV	28,975	81,681	91,681	138,586	218,141	302,478	323,632	345,876	360,738	360,738	360,738	360,738	360,738
SV	-3,330	-13,607	-2,546	-13,342	-19,387	-33,715	-31,163	-39,151	-47,874	-33,044	-22,066	-10,877	0
CV	78	1,781	10,842	1,171	7,387	8,918	6,857	15,882	9,375	11,263	17,982	13,105	11,359
SPI	0.885	0.833	0.972	0.904	0.911	0.889	0.904	0.887	0.867	0.908	0.939	0.970	1.000
CPI	1.003	1.027	1.138	1.009	1.039	1.034	1.024	1.055	1.031	1.036	1.056	1.039	1.033
SCI	0.888	0.856	1.107	0.912	0.946	0.919	0.925	0.935	0.894	0.941	0.991	1.008	1.033
AD	1	2	3	4	5	6	7	8	9	10	11	12	13
ES	0.885	1.742	2.745	3.716	4.756	5.600	5.881	6.201	6.491	7.183	7.676	8.268	9.000
PD	9	9	9	9	9	9	9	9	9	9	9	9	9
SV(t)	-0.115	-0.258	-0.255	-0.284	-0.244	-0.400	-1.119	-1.799	-2.509	-2.817	-3.324	-3.732	-4.000
SPI(t)	0.885	0.871	0.915	0.929	0.951	0.933	0.840	0.775	0.721	0.718	0.698	0.689	0.692
SCI(t)	0.888	0.894	1.042	0.938	0.988	0.965	0.860	0.817	0.743	0.744	0.737	0.716	0.715
$EAC(t)_{PV1}$	9.083	9.339	9.064	9.333	9.484	9.841	9.777	9.977	10.194	9.824	9.551	9.271	9.000
$EAC(t)_{ED1}$	9.115	9.333	9.083	9.385	9.444	9.669	9.674	9.906	10.194	10.916	11.673	12.362	13.000
$EAC(t)_{ES1}$	9.115	9.258	9.255	9.284	9.244	9.400	10.119	10.799	11.509	11.817	12.324	12.732	13.000
$EAC(t)_{PV2}$	10.169	10.799	9.257	9.959	9.878	10.129	9.959	10.149	10.377	9.908	9.586	9.280	9.000
$EAC(t)_{ED2}$	10.169	10.799	9.257	9.959	9.878	10.129	9.959	10.149	10.377	11.008	11.717	12.373	13.000
$EAC(t)_{ES2}$	10.169	10.334	9.835	9.689	9.461	9.642	10.712	11.611	12.479	12.530	12.897	13.062	13.000
$EAC(t)_{PV3}$	10.138	10.516	8.131	9.866	9.511	9.793	9.725	9.623	10.066	9.567	9.077	8.932	8.717
$EAC(t)_{ED3}$	10.141	10.569	8.496	9.903	9.697	9.992	9.890	10.038	10.336	10.974	11.679	12.359	13.000
$EAC(t)_{ES3}$	10.141	10.116	9.003	9.636	9.295	9.522	10.625	11.424	12.375	12.443	12.796	13.022	13.000
TCSPI	1.014	1.048	1.014	1.077	1.111	1.223							
TCSPI(t)	1.014	1.037	1.042	1.057	1.061	1.133							
TCSPI - 11					0.741	0.734	0.669	0.635	0.597				
TCSPI(t) - 11					0.707	0.680	0.780	0.933	1.255				
TCSPI - 13								0.381	0.299	0.305	0.336	0.362	
TCSPI(t) - 13								0.560	0.627	0.606	0.662	0.732	

Table 3.3 Detailed information for project 2 "Link lines" (cost in thousands of €)

	Sep-03	Oct-03	Nov-03	Dec-03	Jan-04	Feb-04	Mar-04	Apr-04	May-04	Jun-04	Jul-04	Aug-04
AC	344	452	796	1,056	1,562	1,922	2,256	2,451	2,676	2,925	3,138	3,247
EV	325	427	735	1,025	1,453	1,774	2,024	2,190	2,356	2,565	2,735	2,875
PV	375	525	850	1,355	1,768	2,125	2,452	2,625	2,875	2,875	2,875	2,875
SV	-50	-98	-115	-330	-315	-351	-428	-435	-519	-310	-140	0
CV	-19	-25	-61	-31	-109	-148	-232	-261	-320	-360	-403	-372
SPI	0.867	0.813	0.865	0.756	0.822	0.835	0.825	0.834	0.819	0.892	0.951	1.000
CPI	0.945	0.945	0.923	0.971	0.930	0.923	0.897	0.894	0.880	0.877	0.872	0.885
SCI	0.819	0.768	0.798	0.734	0.764	0.771	0.741	0.745	0.721	0.782	0.829	0.885
AD	1	2	3	4	5	6	7	8	9	10	11	12
ES	0.867	1.347	2.646	3.347	4.237	5.017	5.717	6.199	6.706	7.653	8.440	9.000
PD	9	9	9	9	9	9	9	9	9	9	9	9
SV(t)	-0.133	-0.653	-0.354	-0.653	-0.763	-0.983	-1.283	-1.801	-2.294	-2.347	-2.560	-3.000
SPI(t)	0.867	0.674	0.882	0.837	0.847	0.836	0.817	0.775	0.745	0.765	0.767	0.750
SCI(t)	0.819	0.636	0.814	0.812	0.788	0.772	0.733	0.692	0.656	0.671	0.669	0.664
$EAC(t)_{PV1}$	9.157	9.307	9.360	10.033	9.986	10.099	10.340	10.362	10.625	9.970	9.438	9.000
$EAC(t)_{ED1}$	9.133	9.373	9.406	9.974	9.891	9.991	10.222	10.326	10.625	11.078	11.536	12.000
$EAC(t)_{ES1}$	9.133	9.653	9.354	9.653	9.763	9.983	10.283	10.801	11.294	11.347	11.560	12.000
$EAC(t)_{PV2}$	10.385	11.066	10.408	11.898	10.951	10.781	10.903	10.788	10.983	10.088	9.461	9.000
$EAC(t)_{ED2}$	10.385	11.066	10.408	11.898	10.951	10.781	10.903	10.788	10.983	11.209	11.563	12.000
$EAC(t)_{ES2}$	10.381	13.363	10.204	10.756	10.621	10.763	11.020	11.615	12.079	11.760	11.730	12.000
$EAC(t)_{PV3}$	10.992	11.713	11.272	12.257	11.773	11.680	12.153	12.073	12.474	11.504	10.855	10.165
$EAC(t)_{ED3}$	10.933	11.596	11.023	12.136	11.398	11.180	11.351	11.120	11.252	11.378	11.646	12.000
$EAC(t)_{ES3}$	10.929	14.028	10.802	10.960	11.042	11.161	11.481	12.046	12.497	12.007	11.837	12.000
TCSPI	1.017	1.053	1.068	1.195	1.223	1.330						
TCSPI(t)	1.017	1.093	1.059	1.131	1.191	1.328						
TCSPI - 11					0.815	0.798	0.805	0.775	0.812			
TCSPI(t) - 11					0.794	0.797	0.821	0.934	1.147			
TCSPI - 12								0.581	0.542	0.539	0.536	
TCSPI(t) - 12								0.700	0.765	0.674	0.560	

Table 3.4 Detailed information for project 3 "Transfer platform" (cost in thousands of €)

	Mar-03	Apr-03	May-03	Jun-03	Jul-03	Aug-03	Sep-03	Oct-03	Nov-03	Dec-03
AC	35	95	174	412	623	754	874	932	952	952
EV	36	93	169	402	597	735	839	887	906	906
PV	34	87	157	373	549	673	798	842	876	906
SV	2	6	12	29	48	62	41	45	30	0
CV	1	-2	-5	-10	-26	-19	-35	-45	-46	-46
SPI	1.059	1.069	1.076	1.078	1.087	1.092	1.051	1.053	1.034	1.000
CPI	1.029	0.979	0.971	0.976	0.958	0.975	0.960	0.952	0.952	
SCI	1.089	1.046	1.046	1.052	1.042	1.065	1.009	1.003	0.984	
AD	1	2	3	4	5	6	7	8	9	10
ES	1.038	2.086	3.056	4.165	5.387	6.496	7.932	9.367	10.000	
PD	10	10	10	10	10	10	10	10	10	
SV(t)	0.038	0.086	0.056	0.165	0.387	0.496	0.912	1.367	1.000	
SPI(t)	1.038	1.043	1.019	1.041	1.077	1.083	1.133	1.171	1.111	
SCI(t)	1.067	1.021	0.989	1.016	1.032	1.055	1.088	1.114	1.057	
$EAC(t)_{PV1}$	9.978	9.934	9.868	9.680	9.470	9.316	9.547	9.503	9.669	
$EAC(t)_{ED1}$	9.941	9.862	9.771	9.689	9.563	9.447	9.640	9.572	9.692	
$EAC(t)_{ES1}$	9.962	9.914	9.944	9.835	9.613	9.504	9.068	8.633	9.000	
$EAC(t)_{PV2}$	9.444	9.355	9.290	9.279	9.196	9.156	9.511	9.493	9.669	
$EAC(t)_{ED2}$	9.444	9.355	9.290	9.279	9.196	9.156	9.511	9.493	9.669	
$EAC(t)_{ES2}$	9.636	9.589	9.818	9.604	9.281	9.236	8.825	8.541	9.000	
$EAC(t)_{PV3}$	9.182	9.556	9.565	9.509	9.596	9.393	9.908	9.974	10.160	
$EAC(t)_{ED3}$	9.210	9.513	9.476	9.410	9.379	9.238	9.616	9.568	9.703	
$EAC(t)_{ES3}$	9.396	9.752	10.020	9.744	9.468	9.320	8.901	8.568	9.000	
TCSPI	0.993	0.983	0.967	0.948	0.913	0.862	0.880	0.786	0.692	
TCSPI(t)	0.996	0.989	0.992	0.973	0.923	0.876	0.689	0.317	0.000	

Chapter 4
A Simulation Study

This chapter presents two extensive simulation studies to test the relevance and accuracy of the concepts and metrics introduced in the previous chapters. The first simulation study has been set up to test the accuracy of the three duration methods (planned value method, earned duration method and earned schedule method) as presented in chapter 1 to predict the final duration of a project. The second simulation study tests the relevance of the schedule adherence concept (measured by the p-factor) as presented in chapter 2.

The methodology used is Monte-Carlo simulation to generate activity duration and cost uncertainty in a project network. The literature on project network simulation is rich and widespread, and is praised as well as criticized throughout various research papers. In these simulation models, activity duration/cost variation is generated using often subjective probability distributions without precise accuracy in practical applications. However, the inability of the simulation runs to incorporate the management focus on a corrective action decision making process to bring late running projects back on track, has led to the crumbling credibility of these techniques. Despite the criticism, practitioners as well as academics have used project network models within a general simulation framework to enable the generation of activity duration and cost uncertainties. For a discussion on the (dis)advantages of project network simulation, the reader is referred to Williams (1999). This issue will not be further discussed in this book.

A summary picture of the methodological approach followed throughout the study is given in figure 4.1, and details are explained in the following subsections. Each project network follows the four steps represented by the four boxes in the figure, which can be summarized as follows:

- Project network generation: A large and diverse set of networks, containing up to 4,100 different project networks with a different topological structure, is generated and presented in section 4.1. To that purpose, various network indicators are used to vary the specific topological structure of each project network which serve as diversity measures in the generated test set.

M. Vanhoucke, *Measuring Time*, International Series in Operations Research & Management Science 136, DOI: 10.1007/978-1-4419-1014-1_4,

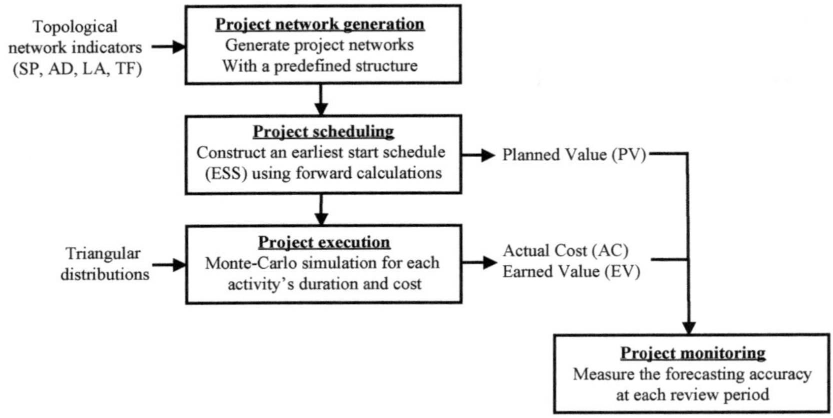

Fig. 4.1 The methodological approach of the simulation experiment

- Project scheduling: Each project network will be transformed into an earliest start Gantt chart, based on the well-known critical path based forward pass calculations. The resulting schedule gives the periodically planned values (PV) for each activity which can be used as the baseline reference point during the EVM simulation study.
- Project execution: The fictitious execution of the project networks is done by simulating random variations in both activity durations and costs under a controlled design. These fictitious project executions result in time-based values for the actual cost (AC) and earned value (EV) which can be translated to EVM metrics during the project monitoring step. These Monte-Carlo simulation runs use triangular distributions tailed to the right or left to generate activities running behind or ahead of schedule. The specific settings of the variability distributions vary throughout the chapter and will be explained later in this chapter, but there is no intention whatsoever to link the generated variability to practical uncertainties or systematic variations such as changes in client requirements, reallocations of staff, over-ambition of technical people, etc. Details of the different Monte-Carlo simulation runs are discussed in sections 4.2 and 4.3.
- Project monitoring: The simulation output consists of the average EVM based performance measurement information obtained in the project execution phase. More precisely, the SPI and SPI(t) indicators and the EAC(t) forecasts measured along the life of the project will be used to calculate their accuracy at the end of the simulation. More details will be provided further in this chapter.

Since any model is a careful abstraction from reality, it inevitably leads to simplifications and often requires restrictive assumptions. Throughout the experiments of the simulation studies of this chapter, it is assumed that:

- Managers do not change their focus on control. Project execution is simulated under different controlled scenarios, but does not take any possible corrective actions into account. Hence, once the project has started, it will be executed according to the different execution scenarios. In chapters 5 and 6, this assumption is relaxed by incorporating corrective management actions to take late running projects back under control.

- Resources are available at all times and can be passed on to succeeding activities at no cost. This implies that activities can be started earlier and/or later than scheduled, depending on the performance of the previous activities.

The outline of this chapter can be summarized as follows. In section 4.1, the project generation process as well as the details of the project network test set used throughout the simulation study are discussed. In section 4.2, a first simulation study to test the forecast accuracy of the earned value predictive methods is discussed in detail. Section 4.3 presents a second simulation study to test the ability of the schedule adherence measure (the p-factor) to improve the accuracy of project duration forecasts.

4.1 Test methodology

In this section, the generation process in order to construct a set of project networks that differ from each other in terms of their topological structure is described in detail. Rather than drawing conclusions for a (limited) set of real life projects, the aim is to generate a large set of project networks that spans the full range of complexity (Elmaghraby and Herroelen, 1980). This guarantees a very large and diverse set of generated networks that can and might occur in practice such that the results of the simulation studies can be generalized. The generation process relies on the project network generator developed by Vanhoucke et al (2008) to generate activity-on-the-node project networks where the set of nodes represents network activities and the set of arcs represents the technological precedence relations between the activities (this format was used for the example project network of figure 1.7). These authors have proposed a network generator that allows generating networks with a controlled topological structure. They have proven that their generator is able to generate a set of very diverse networks that differ substantially from each other from a topological structure point of view. Moreover, it has been shown in literature that the structure of a network heavily influences the constructed schedule (Patterson, 1976), the risk for delays (Tavares et al, 1999), the criticality of a network (Tavares et al, 2004) or the computational effort an algorithm needs to schedule a project (see e.g. Elmaghraby and Herroelen (1980) who call the importance for complexity measures of activity networks to estimate the computational requirements for solution procedures). In the simulation experiments, the design and structure of the generated networks are varied and controlled, resulting in 4,100 diverse networks with 30 activities. Details on the calculations of the project network structure are provided in section 4.1.1. For more information about the specific topological structures and the generation process, the reader is referred to Vanhoucke et al (2008). The constructed data set can be downloaded from www.or-as.be/measuringtime.

4.1.1 The project generation

This section briefly reviews the literature on project network generators and the corresponding topological structures of the generated networks, and presents four topological indicators in detail which will be used throughout the remainder of this book.

Various research papers dealing with network generators for project scheduling problems have been published throughout the academic literature. Demeulemeester et al (1993) have developed a random generator for activity-on-the-arc (AoA[1]) networks. These networks are so-called strongly random since they can be generated at random from the space of all feasible networks with a specified number of nodes and arcs. Besides the number of nodes and the number of arcs, no other characteristics can be specified for describing the network topology. Kolisch et al (1995) describe ProGen, a network generator for activity-on-the-node (AoN) networks which takes into account network topology as well as resource-related characteristics. Schwindt (1995) extended ProGen to ProGen/Max which can handle three different types of resource-constrained project scheduling problems with minimal and maximal time lags. Agrawal et al (1996) recognize the importance of the complexity index CI[2] as a measure of network complexity and have developed an activity-on-the-arc network generator DAGEN for which this complexity measure can be set in advance. Tavares (1999) has presented a new generator RiskNet based on the concept of the progressive level by using 6 morphological indicators (see later in this section). Drexl et al (2000) presented a project network generator ProGen/πx based on the project generator ProGen, incorporating numerous extensions of the classical resource-constrained project scheduling problem. Demeulemeester et al (2003) have developed an activity-on-the-node network generator RanGen which is able to generate a large amount of networks with a given order strength (discussed later). Due to an efficient recursive search algorithm, RanGen is able to generate project networks with exact predefined values for different topological structure measures. The network generator does also take the complexity index CI into account. Akkan et al (2005) have presented a constraint logic programming approach for the generation of acyclic directed graphs. Finally, Vanhoucke et al (2008) have adapted RanGen to an alternative RanGen2 network generator which will be used for the generation of the project networks of this book. It is based on the RiskNet generator of Tavares (1999). Neither of the networks generated by the last eight generators can be called strongly random because they do not guarantee that the topology is a random selection from the space of all possible networks which satisfy the specified input parameters.

Next to the generation of project networks, numerous researchers have spent attention on the topological structure of a project network. The topological structure

[1] In an activity-on-the-arc project network, activities are represented by the arcs and events (or milestones) by the nodes of a network. In this book, the activity-on-the-node network representation is used.

[2] CI is used in this chapter as an abbreviation of the complexity index. In chapter 5, CI will be used to refer to the criticality index.

of a network can be calculated in various ways. Probably the best known measure for the topological structure of activity-on-the-arc networks is the coefficient of network complexity (CNC), defined by Pascoe (1966) as the number of arcs over the number of nodes, and redefined by Davies (1974) and Kaimann (1974, 1975). The measure has been adapted for activity-on-the-node problems by Davis (1975) as the number of direct arcs over the number of activities (nodes) and has been used in the network generator ProGen (Kolisch et al, 1995). Since the measure relies totally on the count of the activities and the direct arcs of the network and as it is easy to construct networks with an equal CNC value but a different degree of difficulty, Elmaghraby and Herroelen (1980) questioned the usefulness of the suggested measure. De Reyck and Herroelen (1996) and Herroelen and De Reyck (1999) conclude that the correlation of the CNC with the complexity index CI is responsible for a number of misinterpretations with respect to the explanatory power of the CNC. Indeed, Kolisch et al (1995) and Alvarez-Valdes and Tamarit (1989) had revealed that resource-constrained project scheduling networks become easier with increasing values of the CNC, without considering the underlying effect of the CI. In conclusion, the CNC, by itself, fails to discriminate between easy and hard project networks and can therefore not serve as a good measure for describing the impact of the network topology on the hardness of a project scheduling problem.

Another well-known measure of the topological structure of an AoN network is the order strength, OS (Mastor, 1970), defined as the number of precedence relations (including the transitive[3] ones but not including the arcs connecting the dummy start or end activity) divided by the theoretical maximum number of precedence relations $\frac{n*(n-1)}{2}$, where n denotes the number of non-dummy activities in the network). It is sometimes referred to as the density (Kao and Queyranne, 1982) or the restrictiveness (Thesen, 1977) and equals 1 minus the flexibility ratio (Dar-El, 1973). Herroelen and De Reyck (1999) conclude that the order strength OS, the density, the restrictiveness and the flexibility ratio constitute one and the same complexity measure. Schwindt (1995) uses the order strength in the problem generator ProGen/Max and argues that this measure plays an important role in predicting the difficulty of different resource-constrained project scheduling problems. De Reyck (1995) verified and confirmed the conjecture that the OS outperforms the complexity index CI as a measure of network complexity for the resource-constrained project scheduling problem.

The complexity index CI was originally defined by Bein et al (1992) for two-terminal acyclic activity-on-the-arc networks as the reduction complexity, i.e. the minimum number of node reductions which – along with series and parallel reductions – allow to reduce a two-terminal acyclic network to a single edge. As a consequence, the CI measures the closeness of a network to a series-parallel directed graph. Their approach for computing the reduction complexity consists of two steps. First, they construct the so-called complexity graph by means of a dominator and a reverse-dominator tree. Second, they determine the minimal node cover through

[3] When two direct or immediate precedence relation exist between activities (i, j) and activities (j, k), then there is also an implicit transitive relation between activities (i, k).

the use of the maximum flow procedure by Ford and Fulkerson (1962). De Reyck and Herroelen (1996) adopted the reduction complexity as the definition of the complexity index CI of an activity network and have proven the CI to outperform other popular measures of performance, such as the CNC. Moreover, they also show that the OS, on its turn, outperforms the CI. These studies motivated the construction of an AoN problem generator for networks where both the order strength OS and the complexity index CI can be specified in advance, which has led to the development of the RanGen and RanGen2 generators, used throughout this book.

In this book, the topological structure of an activity-on-the-node network is calculated based on four indicators initially proposed by Tavares et al (1999, 2002) and further developed by Vanhoucke et al (2008). These indicators serve as classifiers of project networks by controlling the design and structure of each individual project network. All indicators have been rescaled and lie between 0 and 1, inclusive, denoting the two extreme structures. The logic behind each indicator is often straightforward, and relies on general topological definitions from the project scheduling literature. Their specific calculations and intuitive meaning are the topic of the current section.

4.1.1.1 Topological definitions

This book relies on the activity-on-the-node project network representation in which the nodes represent activities and the arcs represent direct (or immediate) precedence constraints. In order to be in line with the literature, a project network can be extended with a dummy start node and a dummy end node (see e.g. the dummy nodes 1 and 12 of figure 1.7). Further network characteristics that will be used in the description of the topological network indicators are:

n	Number of non-dummy activities in an AoN network
N	Set of non-dummy project activities, from 1 to n
P_i	Set of immediate predecessors of activity i
S_i	Set of immediate successors of activity i

In the example project network of figure figure 1.7, n = 10 and N = {2, 3, 4, 5, 6, 7, 8, 9, 10, 11}. The set of immediate predecessors is equal to $P_2 = P_3 = P_4 = \emptyset$, $P_5 = \{2\}, P_6 = \{4\}, P_7 = \{6\}, P_8 = \{7\}, P_9 = \{5\}, P_{10} = \{3\}, P_{11} = \{7\}$. Likewise, the set of immediate successors is equal to $S_2 = \{5\}, S_3 = \{10\}, S_4 = \{6\}, S_5 = \{9\}$, $S_6 = \{7\}$, $S_7 = \{8, 11\}$, $S_8 = S_9 = S_{10} = S_{11} = \emptyset$. The calculations of the four indicators rest on a number of straightforward definitions that characterize the project network. The progressive and regressive level of activities in a project network have been defined by Elmaghraby (1977) as follows:

- **Progressive level:** The progressive level of an activity i in a project network is defined as:

$$PL_i = \begin{cases} 1 & \text{if } P_i = \emptyset \\ \max_{j \in P_i} PL_j + 1 & \text{if } P_i \neq \emptyset \end{cases} \tag{4.1}$$

- **Regressive level:** The regressive level of an activity i in a project network is defined as:

$$RL_i = \begin{cases} m & \text{if } S_i = \emptyset \\ \min_{j \in S_i} RL_j - 1 & \text{if } S_i \neq \emptyset \end{cases} \tag{4.2}$$

with m the maximal progressive level, i.e. $m = max_{i \in N} PL_i$.

Based on the definition of the progressive and regressive level in an activity network, the following definitions can be used:

- **Width:** The width w_a of each progressive level $a = 1, \ldots, m$ is defined as the number of activities at that level.
- **Length of an arc:** The length l of an arc (i, j) is equal to the difference between the progressive level of the end node j and the start node i, i.e. $PL_j - PL_i$.
- **Topological float:** The topological float of an activity i is equal to the difference between the regressive level and the progressive level of activity i, i.e. $RL_i - PL_i$.

These five definitions are necessary during the calculation of the four project network indicators as defined below.

4.1.1.2 Topological indicators

A. Serial or parallel indicator SP.

The first indicator, SP $\in$ [0, 1], measures the closeness of a network to a serial or parallel network. More precisely, when SP = 0 then all activities are in parallel, and when SP = 1 then the project network is completely serial. Between these two extreme values, networks can be generated close to a serial or parallel network. Hence, the SP indicator determines the maximal number of levels of the network, defined as the longest chain (in terms of the number of serial activities) in the network.

The SP indicator can be formulated as follows:

$$SP = \begin{cases} 1 & \text{if } n = 1 \\ \frac{m-1}{n-1} & \text{if } n > 1 \end{cases} \tag{4.3}$$

B. Activity distribution AD.

The second indicator, AD $\in$ [0, 1], measures the distribution of project activities along the levels of the project, and hence, takes the width of each progressive level into account. When AD = 0, all levels contain a similar number of activities, and hence, the number of activities is uniformly distributed over all levels. When AD = 1, there is one level with a maximal number of activities, and all other levels contain a single activity.

The AD indicator can be formulated as follows:

$$AD = \begin{cases} 0 & \text{if } m \in \{1,n\} \\ \frac{\alpha_w}{\alpha_{max}} = \frac{\sum_{a=1}^{m} |w_a - \bar{w}|}{2(m-1)(\bar{w}-1)} & \text{if } m \notin \{1,n\} \end{cases} \tag{4.4}$$

Consequently, this indicator measures the distribution of the activities over the progressive levels by calculating the total absolute deviations α_w and α_{max}. α_w measures the total absolute deviation of the activity distribution $w = (w_1, w_2, \ldots, w_m)$ from the average deviation $\bar{w} = n/m$ as $\alpha_w = \sum_{a=1}^{m} |w_a - \bar{w}|$.

α_{max} determines the maximal value of α_w for a network with n activities and m progressive levels. α_{max} corresponds to a network for which $m-1$ progressive levels have a width w_a of 1, and one progressive level has a width w_a of $n-(m-1)$. The value of can be calculated as $\alpha_{max} = (m-1)(\bar{w}-1) + (n-m+1-\bar{w})$.

The first term calculates the absolute deviation between $w_a = 1$ and the average width $\bar{w}$ for $m-1$ progressive levels. The second term calculates the difference between $w_a = n-(m-1)$ and $\bar{w}$ for the remaining progressive level. The formula for $m \neq 1$ can be simplified to $\alpha_{max} = 2(m-1)(\bar{w}-1)$, resulting in the AD indicator defined above. This indicator equals 1 when $\alpha_w = \alpha_{max}$. At the other extreme, the indicator has a value of 0 when the activities are uniformly distributed over the progressive levels, i.e. $w_a = \bar{w} = n/m$ (for $a = 1, \ldots, m$).

C. Length of arcs LA.

The third indicator, LA ∈ [0, 1], measures the length of each precedence relation (i, j) in the network as the difference between the level of the end activity j and the level of the start activity i. When LA equals 0, the network has many precedence relations between two activities on levels far from each other. Hence, the activity can be shifted further in the network. When LA equals 1, many precedence relations have a length of one, resulting in activities with immediate successors on the next level of the network, and hence little freedom to shift.

In order to define the LA indicator, a parameter n'_l is defined as the number of arcs in the network with length l (note that the length l of an arc can vary between 1 (short) and $m-1$ (long)). Based on this parameter, the LA $\in [0,1]$ indicator measures the presence of short (i.e. with a length $l = 1$) immediate precedence relations and can be defined as follows:

$$LA = \begin{cases} 1 & \text{if } D = n - w_1 \\ \frac{n'_1 - n + w_1}{D - n + w_1} & \text{if } D > n - w_1 \end{cases} \tag{4.5}$$

where D stands for the maximal number of short ($l = 1$) precedence relations in a network, given the width of each level, i.e. $D = \sum_{a=1}^{m-1} w_a w_{a+1}$.

D. Topological float TF.

The last indicator, TF $\in$ [0, 1], measures the topological float of a precedence relation as the number of levels each activity can shift without violating the maximal level of the network (as defined by SP). Hence, TF = 0 when the network structure is 100% dense and no activities can be shifted within its structure with a given SP value. A network with TF = 1 consists of one chain of activities without topological float (they define the maximal level and hence, the SP value) while the remaining activities have a maximal float value (which equals the maximal level, defined by SP, minus 1).

$$TF = \begin{cases} 0 & \text{if } m \in \{1,n\} \\ \frac{\sum_{i=1}^{n} RL_i - PL_i}{(m-1)(n-m)} & \text{if } m \notin \{1,n\} \end{cases} \tag{4.6}$$

4.1.1.3 Illustrative examples

Figure 4.2 shows the example network of figure 1.7 containing 10 non-dummy activities divided among four (progressive or regressive) levels, indicated by the zones between the vertical dashed lines. The start and end dummy activities and their corresponding outgoing and incoming precedence arcs are not taken into account for the topological structure calculation since they do not represent real activities (i.e. with a real duration, cost, risk, etc.). The values for the different indicators are calculated below.

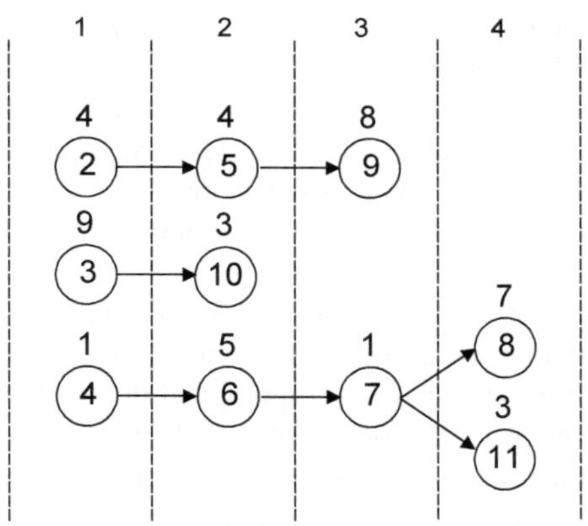

Fig. 4.2 The example project network of figure 1.7 without start and end dummy

Table 4.1 reports the values for the progressive and regressive level for each activity. Based on this information, the values for all four indicators can be calculated. In this example, $n = 10$, $m = 4$, $w_1 = 3$, $w_2 = 3$, $w_3 = 2$, $w_4 = 2$ and $n'_1 = 7$ and consequently $D = 3*3+3*2+2*2 = 19$.

Table 4.1 Progressive and regressive level of each activity in figure 4.2

i	2	3	4	5	6	7	8	9	10	11
PL_i	1	1	1	2	2	3	4	3	2	4
RL_i	2	3	1	3	2	3	4	4	4	4

The values for the indicators can now be calculated by filling in all the needed parameters, as follows:

- $SP = \frac{(4-1)}{(10-1)} = 0.33$
- $AD = \frac{(|3-2.5|+|3-2.5|+|2-2.5|+|2-2.5|}{2(4-1)(2.5-1)} = \frac{2}{9} = 0.22$
- $LA = \frac{(7-10+3)}{(19-10+3)} = 0$
- $TF = \frac{((2-1)+(3-1)+(1-1)+(3-2)+(2-2)+(3-3)+(4-4)+(4-3)+(4-2)+(4-4))}{(4-1)(10-4)} = \frac{7}{18} = 0.39$

Figure 4.3 displays 9 example networks with 9 non-dummy activities with an SP value of 0.25, resulting in projects with three levels. It is intuitively clear that all networks are closer to a parallel network (1 level) than to a serial network (9 levels). A complete parallel network (SP = 0) would result in 9 activities without precedence relations (i.e. only 1 level) while a complete serial network (SP = 1) would result in 9 activities on a chain (9 levels). The first (second; third) row of the figure displays networks with an AD value of 0 (0.25; 1). The lower the AD value, the more balanced the individual activities over the three levels of the project. The first (second; third) column of the figure displays networks with an LA value of 0 (0.5; 1). As explained in the text of this paper, LA measures short precedence relations in the network and hence, the more short precedence relations in the network, the higher the value for LA.

The TF values for the networks of figure 4.3 (from (a) to (i)) equal 0, 0, 0, 0.25, 0, 0, 1, 0.5 and 0. In the figure, all networks contain only short precedence relations between sets of activities. However, the topological float (TF) of a network is determined by *all* the precedence relations (both the short ones as measured by LA but also the long precedence relations). As an example, figure 4.4 displays the network (h) of figure 4.3 with some additional precedence relations. Although the LA value remains unchanged, the extra precedence relations result in a change of the TF value from 0.5 to 0.25 and hence, in a change of the network structure.

For more information about the specific calculations of the indicators and their use in various research settings, the reader is referred to Vanhoucke et al (2008). In this paper, the authors refer to SP as the I_2 indicator, AD as the I_3 indicator, LA as the I_4 indicator and TF as the I_6 indicator. More information on similar network

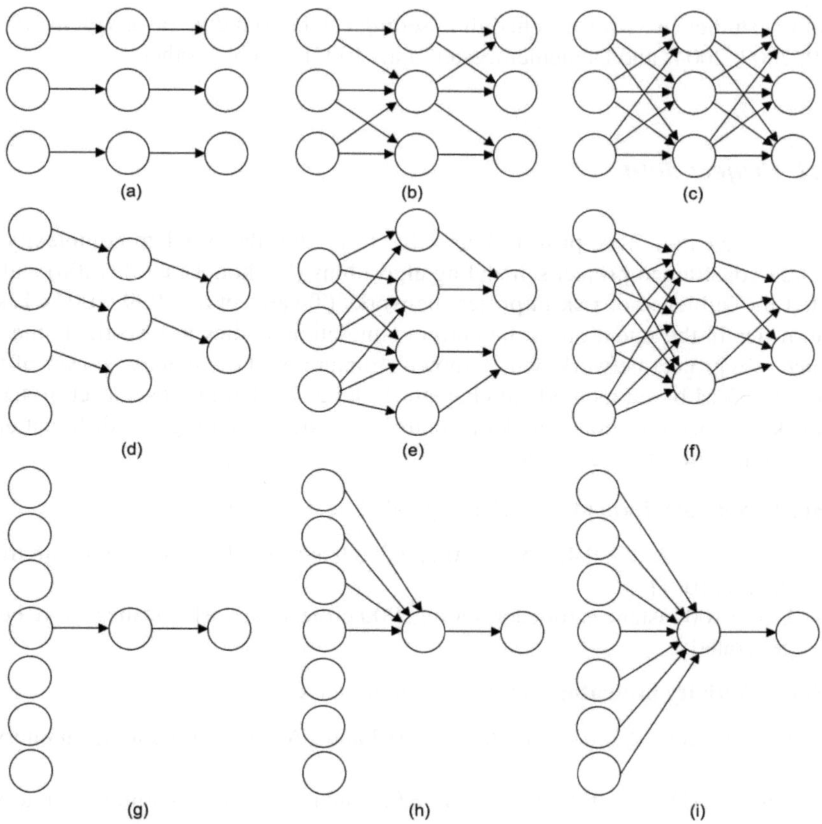

Fig. 4.3 9 example networks with an SP value of 0.25

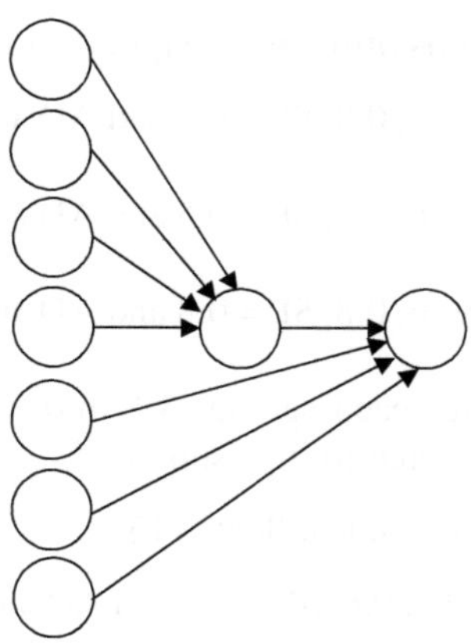

Fig. 4.4 Network (h) of figure 4.3 with additional precedence relations

topology studies in a project scheduling setting can be found in Tavares et al (1998, 1999, 2002, 2004) and Demeulemeester et al (2003), amongst others.

4.1.2 Project data

The four indicators have proven their value to predict the problem complexity of resource-constrained project scheduling algorithms (Vanhoucke et al, 2008) and to predict the influence of risk in project networks (Tavares et al, 1999, 2004). Likewise, it lies in the intention of the current simulation studies to investigate the influence of a project network with a known structure on the predictive power of the proposed EVM based methods of chapters 1 and 2. To that purpose, a set of 4,100 networks has been generated under a controlled design, resulting in 4 different data sets, with the following characteristics:

Set 1: Network indicator: serial or parallel network (SP)

SP = 0.1; 0.2; 0.3; 0.4; 0.5; 0.6; 0.7; 0.8; 0.9 and AD, LA and TF random from interval [0, 1]
→ Using 100 instances for each setting, 900 project network instances have been generated.

Set 2: Activity indicator: activity distribution (AD)

Set 2.1: AD = 0.2; 0.4; 0.6; 0.8, SP = 0.2 and LA and TF random from interval [0, 1]
Set 2.2: AD = 0.2; 0.4; 0.6; 0.8, SP = 0.5 and LA and TF random from interval [0, 1]
→ Using 100 instances for each setting, 2 * 400 = 800 project network instances have been generated for this subset.

Set 3: Precedence relations indicator: length of arcs (LA)

Set 3.1: LA = 0.2; 0.4; 0.6; 0.8, SP = 0.2 and AD and TF random from interval [0, 1]
Set 3.2: LA = 0.2; 0.4; 0.6; 0.8, SP = 0.5 and AD and TF random from interval [0, 1]
Set 3.3: LA = 0.2; 0.4; 0.6; 0.8, SP = 0.8 and AD and TF random from interval [0, 1]
→ Using 100 instances for each setting, 3 * 400 = 1,200 project network instances have been generated for this subset.

Set 4: Float indicator: topological float (TF)

Set 4.1: TF = 0.2; 0.4; 0.6; 0.8, SP = 0.2 and AD and LA random from interval [0, 1]
Set 4.2: TF = 0.2; 0.4; 0.6; 0.8, SP = 0.5 and AD and LA random from interval [0, 1]

Set 4.3: TF = 0.2; 0.4; 0.6; 0.8, SP = 0.8 and AD and LA random from interval [0, 1]

→ Using 100 instances for each setting, 3 * 400 = 1,200 project network instances have been generated for this subset.

Each set measures the network structure with the intuitive logic as explained above. This data set can be downloaded in the ProTrack format (see chapter 7) from www.or-as.be/measuringtime. Each network contains 30 non-dummy activities.

4.2 Simulation 1: A forecast accuracy study

4.2.1 Simulation model

The simulation step of figure 4.1 generates for each baseline project schedule 100 real life project executions resulting in an average project performance that might differ from the original baseline performance. Project execution is submitted to artificial uncertainty by introducing variation in activity durations and costs through 9 simulation scenarios. More precisely, the average project performance is carefully controlled by the introduction of a separate variation on critical and non-critical activities. Table 4.2 displays the 9 simulation scenarios of the computational experiment on all 4,100 generated networks, and reads as follows:

- **Critical versus non-critical activities:** A distinction has been made between critical and non-critical activities. Each (critical and non-critical) activity can have an actual (i.e. simulated) duration which is smaller than (-), equal to (0) or larger than (+) its corresponding planned duration.
- **Actual project performance:** The actual (i.e. simulated) project performance at completion is measured by comparing the real project duration RD with the planned duration PD. Hence, each column reflects a known schedule condition as follows:
  - 1st column: early project completion (RD $<$ PD),
  - 2nd column: on schedule (RD = PD),
  - 3rd column: late project completion (RD $>$ PD).
- **Measured project performance:** The $\overline{\text{SPI(t)}}$ index is used to calculate the average reported project duration performance, measured as the average SPI(t) value calculated for each review period during the entire project execution.

The simulation runs follow the design of table 4.2 and allow validating the three predictive techniques of chapter 1 and comparing their relative performance. The $\overline{\text{SPI(t)}}$ indicator, which is used as an average predictor of the overall project performance throughout the entire project life cycle, plays a central role and might act as a (correct or misleading) project performance warning signal of project performance. As an example, scenario 3 measures – on the average over all reporting periods –

Table 4.2 9 simulation scenarios for our computational tests

Non-critical activities	Critical activities: -	Critical activities: 0	Critical activities: +
-	**1** $\overline{\text{SPI(t)}} > 1$ RD < PD	**4** $\overline{\text{SPI(t)}} > 1$ RD = PD	**7** $\overline{\text{SPI(t)}} > 1$ RD > PD
0	**2** $\overline{\text{SPI(t)}} > 1$ RD < PD	**5** $\overline{\text{SPI(t)}} = 1$ RD = PD	**8** $\overline{\text{SPI(t)}} < 1$ RD > PD
+	**3** $\overline{\text{SPI(t)}} < 1$ RD < PD	**6** $\overline{\text{SPI(t)}} < 1$ RD = PD	**9** $\overline{\text{SPI(t)}} < 1$ RD > PD

a project delay ($\overline{\text{SPI(t)}} < 1$), but the project finishes earlier than expected (RD < PD). Hence, scenario 3 is set up to simulate a misleading performance indication that the project will end later than originally planned. Scenario 8 measures an average project delay, which is a correct project performance warning signal since RD > PD. The 9 scenarios can be interpreted as follows:

Scenario 1: A correct average project ahead performance indication since the project finishes ahead of schedule
Scenario 2: A correct average project ahead performance indication since the project finishes ahead of schedule
Scenario 3: An average project delay performance indication but the project finishes ahead of schedule
Scenario 4: An average project ahead performance indication but the project finishes on time
Scenario 5: A 100% on time performance indication and real project execution
Scenario 6: An average project delay performance indication but the project finishes on time
Scenario 7: An average project ahead performance indication but the project finishes behind schedule
Scenario 8: A correct average project delay performance indication since the project finishes behind schedule
Scenario 9: A correct average project delay performance indication since the project finishes behind schedule

All EVM metrics presented in the previous chapters make no distinction between critical and non-critical activities and suffer from the fact that all activities have an equal weight in the total earned value calculations. Therefore, a distinction is made between critical and non-critical activities throughout the simulation runs in order to test the potential false or misleading warning signal EVM might observe in predicting the final project duration (see section 4.3). Note that four scenarios (1, 2, 8 and 9) give a correct warning signal during the execution of the project, and four scenarios (3, 4, 6 and 7) give a false warning signal. One scenario (5) assumes no uncertainty at all, i.e. all planned durations equal the actual durations. Throughout all simulations, the random variations in both (critical and non-critical) activity durations and costs are based on triangular distributions tailed to the right (activity delays) or to the left (activities ahead of schedule). The triangular distribution is a continuous probability distribution $f(x)$ with lower limit a, mode c and upper limit b, i.e.

$$f(x|a,b,c) = \begin{cases} \frac{2(x-a)}{(b-a)(c-a)} & \text{for } a \leq x \leq c \\ \frac{2(b-x)}{(b-a)(b-c)} & \text{for } c < x \leq b \end{cases} \tag{4.7}$$

The input parameters of a triangular distribution, i.e. a lower limit a, a mode c and an upper limit b is carefully selected and obviously depend on the scenario to simulate. As an example, scenario 1 (9) assumes activities finishing earlier (later) than planned, which can be easily simulated by means of triangular distribution tailed to the left (right) to simulate the earliness (lateness). Other scenarios contain a mix of left-tailed and right-tailed triangular distributions for the project activities, depending on the settings and characteristics of the scenario. However, not all scenarios are straightforward to simulate due to the specific character of each scenario. As an example, the simulation of scenario 3 (i.e. simulate a project finishing ahead of schedule but the EVM performance indicators measure the opposite) depends heavily on the structure of the network. Indeed, the duration of the critical activities needs to be decreased, but the non-critical activities need to be increased to guarantee that the EVM performance indicators measure an average SPI(t) $<$ 1 along the life of the project, although the project finishes ahead of schedule. To obtain this scenario, the duration of the non-critical activities needs to be increased as much as possible within their activity slack (resulting in an average SPI(t) $<$ 1). A project network consisting of many serial activities has only a few non-critical activities, and hence, a careful selection of simulation values for the real activity durations is necessary. More precisely, only a few critical activities will be decreased to a very small extent (such that RD $<$ PD) while the few non-critical activities need to be increased as much as possible within their slack (such that the SPI(t) value is, on the average, smaller than 1). Therefore, each scenario has been simulated under strict conditions and hence, comparison *between* scenarios is often of little value.

The cost deviations are assumed to deviate from the original budget (BAC per activity) in correlation with the duration deviation. In doing so, it is assumed that the cost is expressed in € per man-hour and hence, deviations in activity duration have an immediate effect on the cost, due to an increase or a decrease in the total amount of man-hours to finish the particular activity. Although this reflects many real life situations in project management, one can consider other settings where the cost deviation has another relation (or no relation) to the duration of an activity. However, the focus of this book is on the prediction of a project's final duration, and not on cost. The SCI and SCI(t) metrics are only used to adapt the planned duration of work remaining (PDWR, see chapter 1) when one assumes that future performance is likely to follow the current SCI or SCI(t) trends, which only makes sense when the cost is correlated with duration performance.

The remainder of section 4.2 presents extensive results for various simulation runs and is divided into three subsections, inspired by the three criteria proposed by Covach et al (1981) for evaluating the performance of EAC methods, i.e. accuracy, timeliness and stability.

- **Forecast accuracy:** Section 4.2.2 evaluates the overall forecast accuracy of the three methods (planned value, earned duration and earned schedule) for the 9 proposed scenarios.
- **Timeliness:** Section 4.2.3 analyzes the behavior of the forecasts along the completion stage of the generated projects, and hence, measures whether the forecasting methods are capable of producing reliable results in the early, middle and late stages of the project life cycle.
- **Stability:** In section 4.2.4, the influence of the SP indicator on the forecast accuracy is discussed in order to measure the robustness or stability of the forecasting measures for different network structures. Note that this approach deviates from the original definition of stability of Covach et al (1981) that refers to the stability of forecasting methods over the different review periods, and not over the different structures of the network.

In order to evaluate the forecasting measures and to determine the forecast accuracy of each technique, two straightforward measures are calculated. The measures calculate the average deviations between the total project durations predicted during project execution (EAC(t)) and the final real project duration (RD) observed after the finish of the project. Obviously, the lower their value, the more accurate the average duration prediction. The two measures are:

- Mean Percentage Error (MPE): $\frac{1}{T}\sum_{time=1}^{T}\frac{EAC(t)^{time}-RD}{RD}*100$
- Mean Absolute Percentage Error (MAPE): $\frac{1}{T}\sum_{time=1}^{T}\frac{|EAC(t)^{time}-RD|}{RD}*100$

where T is used to refer to the total number of reporting periods over the complete project horizon and $EAC(t)^{time}$ is used to denote the estimated duration at completion in reporting period time (time = 1, 2, ..., T) (more precisely, at each reporting period, a corresponding duration forecast $EAC(t)^{time}$ is calculated).

The calculation of the MAPE has been illustrated earlier for the example project of chapter 1 with a baseline project duration PD = 16 and a real life duration RD = 18. The EAC(t) values along the life of the project, from time period 1 to 18, are shown in the graphs of figure 1.11 (each period time shows an $EAC(t)^{time}$ value). The periodic forecast accuracy is given in table 1.5 and the average value at the bottom row of this table is equal to the MAPE as defined here. The MPE can be calculated in a similar way, but, unlike the MAPE, positive and/or negative values are possible to measure over- and/or underestimations of the final project duration, respectively.

4.2.2 The forecast accuracy under 9 scenarios

The results obtained from the simulation runs under the 9 scenarios need to be interpreted with care. Recall that scenarios 1, 2, 8 and 9 are said to be "correct" scenarios since the average project performance metric (measured by $\overline{\text{SPI}(t)}$) gives a correct warning signal for the final real project duration RD. The main results should focus on these scenarios since they represent normal project behavior. However, scenarios 3, 4, 6 and 7 are set up as rather "extreme" scenarios since the activity duration and cost deviations are simulated to force a misleading EVM performance warning signal. These scenarios need to be considered as exceptional scenarios and consequently, are less suitable for drawing general conclusions.

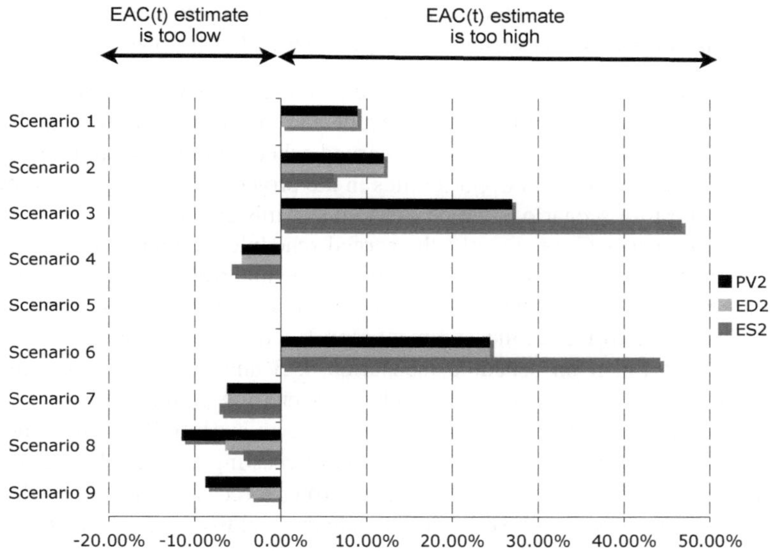

Fig. 4.5 The forecast accuracy (MPE) for the 9 scenarios

In the remainder of this chapter, each EAC(t) forecasting metric has been abbreviated to e.g. $\text{EAC}(t)_{\text{PV1}}$ = PV1. Figure 4.5 illustrates the MPE values for the PV2, ED2 and ES2 methods (where it is assumed that the future performance follows the current SPI or SPI(t) trend). The figure shows that, on average, the earned schedule method outperforms the two other predictive methods under normal project circumstances (i.e. the correct scenarios 1, 2, 8 and 9) and hence can be considered as the most accurate forecasting method. However, this method is outperformed by the other two methods for the misleading scenarios. Detailed results and a critical discussion will be given below in the following tables of this section.

Table 4.3 The forecast accuracy (MAPE) of the three methods for the 9 scenarios

Scenario	**PV1**	**PV2**	**PV3**	**ED1**	**ED2**	**ED3**	**ES1**	**ES2**	**ES3**
1	35.43	22.97	33.10	40.24	22.97	20.86	**32.16**	**12.51**	**19.61**
2	31.29	22.52	24.86	33.84	22.52	21.13	**28.25**	**13.72**	**15.37**
3	8.44	**27.04**	269.94	**6.52**	**27.04**	**181.83**	8.18	46.68	288.48
4	1.88	**4.52**	23.81	**1.36**	**4.52**	**13.63**	1.75	5.67	14.40
5	**0.00**	**0.00**	**0.00**	**0.00**	**0.00**	**0.00**	**0.00**	**0.00**	**0.00**
6	6.13	**24.40**	268.61	**4.25**	**24.40**	**177.87**	6.14	44.14	285.69
7	3.73	6.35	33.17	**2.99**	**6.27**	**18.78**	3.16	7.15	19.18
8	21.63	19.07	22.00	17.15	14.64	17.85	**15.56**	**7.76**	**11.75**
9	21.16	17.73	26.69	16.73	13.16	19.73	**14.95**	**5.73**	**18.11**

Table 4.3 displays the mean absolute percentage error (MAPE) for all the scenarios for the three proposed methods. The forecast accuracy results differ according to the settings of each scenario, and can be summarized as follows.

- Projects that finish ahead of schedule (scenarios 1, 2 and 3): the earned schedule method outperforms both the planned value and the earned duration method for scenarios 1 and 2. In these cases, the earned schedule method can be used as a reliable indicator to detect opportunities in the project. Note that the ED method performs best for scenario 3, which is set up as a misleading project performance indicator scenario. Consequently, the earned schedule method performs best under normal circumstances (i.e. when correct schedule performance indicator values are measured along the life of the project), and performs less good in case the schedule performance measurement reports a misleading warning signal.
- Projects that finish on schedule (scenarios 4, 5 and 6): The earned duration method outperforms the other methods for scenarios 4 and 6. However, these scenarios have especially been built to generate an average SPI(t) indicator that gives a false warning signal and hence, the forecasting metrics should be influenced by this false indicator, resulting in wrong forecasts. The earned duration method suffers less than the earned schedule method from this misleading warning signals, since the SPI indicator used in the ED method tends to go to 1 at the end of the project, decreasing the error of the false warning signal.
- Projects that finish behind schedule (scenarios 7, 8 and 9): the earned schedule method outperforms the other methods, which means it can be used to detect problems in projects. However, in scenario 7, the earned duration has the best performance. This is also a scenario for which the SPI(t) indicator gives a false warning signal.

Overall, the table shows that all methods perform best when the performance factor is equal to SPI (PV2 and ED2) or SPI(t) (ES2) for the normal simulation scenarios. Consequently, the most accurate results are obtained when the future expected performance is assumed to be in line with the current observed project performance. None of the forecasting metrics perform very well when the performance measure is equal to 1 (PV1, ED1 and ES1) or equal to SCI (PV3, ED3 and ES3) under normal project behavior. Consequently, correcting the future expected project performance, or correcting the forecasting metrics with cost information (the SCI or SCI(t) metrics are used in the denominator) does not lead to reliable results for the three methods, and should be excluded. Anbari (2003) points out that the time estimate at completion adjusted for cost performance may provide a better indication of estimated time at completion, when adherence to budget is critical to the organization. He points out that additional time may be needed to bring the project back on budget (by reducing resources applied to the project, taking additional time to find better prices for equipment and material, and similar actions). In this book, the impact of cost performance on the schedule is not explicitly taken into account.

It should be noted that ED2 and PV2 report exactly the same forecast accuracy for scenarios 1 to 6. The formulas in chapter 1 clearly show that, in case the real project duration RD at the reporting periods is smaller than the PD, the $EAC(t)_{PV2}$ and $EAC(t)_{ED2}$ are exactly the same and equal to $\frac{PD}{SPI}$. However, from the moment the real project duration RD exceeds the planned duration PD, both predictive methods will produce different results for reporting periods later than the planned project duration PD. Also note that the significance of all differences has been verified with a non-parametric test in SPSS. All differences as indicated in the table (the best performing method has been indicated in bold) were statistically significant (α = 0.05).

Table 4.4 The forecast accuracy (MPE) of the three methods for the 9 scenarios

Scenario	**PV1**	**PV2**	**PV3**	**ED1**	**ED2**	**ED3**	**ES1**	**ES2**	**ES3**
1	33.81	8.94	-29.25	35.65	8.94	-11.04	31.86	-0.03	-16.72
2	29.98	11.96	-4.39	30.36	11.96	3.81	27.87	6.12	-1.48
3	8.39	26.95	269.94	6.46	26.95	181.79	8.18	46.65	288.47
4	-1.88	-4.52	-23.81	-1.36	-4.52	-13.63	-1.75	-5.67	-14.40
5	0.00	0.00	0.00	0.00	0.00	0.00	0.00	0.00	0.00
6	6.13	24.40	268.61	4.25	24.40	177.87	6.14	44.14	285.69
7	-3.68	-6.23	-33.16	-2.95	-6.15	-18.76	-3.16	-7.09	-19.18
8	-21.59	-11.50	3.09	-16.92	-6.46	1.08	-15.56	-4.31	2.95
9	-21.12	-8.76	23.49	-16.49	-3.51	12.86	-14.92	-0.22	17.10

Table 4.4 shows the MPE for the 9 scenarios and confirms the reported results of table 4.3 that the earned schedule method (ES) outperforms on average both the planned value (PV) and earned duration (ED) methods. The table is particularly interesting to discuss the influence of misleading performance warning signals (scenarios 3, 4, 6 and 7) when compared to the correct scenarios 1, 2, 8 and 9. The table shows that misleading EVM information and false project performance indicators have an immediate effect on the accuracy of the EAC(t) measures, with the ES method showing the lowest accuracy. As an example, scenario 3 reports a project delay during its progress, although the project ultimately finishes early. Hence, the forecasts clearly show an overestimation (due to the false SPI and SPI(t) warnings). In these cases, the forecasts are no longer reliable, which explains the low performance of the ES method. Consequently, the SPI(t) indicator can be considered as a reliable measure for the project performance and hence, the forecast accuracy is

eventually determined by the quality of the reported SPI(t) values along the life of the project. In case the SPI(t) reports false warning signals (cf. scenarios 3, 4, 6 and 7), the forecast accuracy suffers from this misleading project performance indication, resulting in a poor predictive quality of the EAC(t) for the ES method. Since the SPI indicator is less reliable compared to the SPI(t) (certainly at the late stage of the project, where the SPI indicator tends to go to one, regardless of the real project performance), the forecast accuracy is more a random guess having an average forecast accuracy which does not vary as much between the 9 scenarios as for the ES method. Hence, the difference between correct SPI reports (scenarios 1, 2, 8 and 9) and false SPI reports (scenarios 3, 4, 6 and 7) is less outspoken than for the SPI(t) indicator in the ES method.

In the remainder of this book, results will often be reported for the PV2, ED2 and ES2 methods under normal project behavior, since these methods perform relatively better than the other versions of the same predictive methods.

4.2.3 The forecast accuracy and the completion stage of work

In order to test the relation between the forecast accuracy and the completion stage of the project, this section reports results for additional simulation runs that measure the accuracy of time forecasts as a function of the completion stage of the project. Lipke (2003) has shown that the classic schedule indicators (SV and SPI) are unreliable as project duration forecasting indicators since they show a strange behavior over the final third of the project. This problem is overcome by the earned schedule concept which behaves correctly over the complete project horizon (see e.g. figure 1.5 or the examples in chapter 3). In order to investigate the effect of the behavior of SPI and SPI(t) on the forecasting measures, the overall (duration) performance along the completion stage of the project (expressed in their percentage completed $\frac{EV}{BAC}$) is measured. The project horizon is divided in three stages (early, middle and late). Each stage measures the current project performance based on earned value information obtained in that stage, and forecasts the final project duration. The early stage consists of the first 30% completed, the middle stage contains all work between 30% and 70% completed and the late stage contains the final 30% work completed. The review periods during project tracking are divided in steps of 10% completed, such that the early, middle and late stage contain, respectively, 3, 4 and 3 review periods as given in table 4.5.

Table 4.5 The simulation scenarios with different work completion stages

	Percentage completed:
1. Early stage	0% - 10%, 10% - 20%, 20% - 30%
2. Middle stage	30% - 40%, 40% - 50%, 50% - 60%, 60% - 70%
3. Late stage	70% - 80%, 80% - 90%, 90% - 100%

In this simulation run, the behavior of the three schedule forecasting methods is analyzed along the completion stage of the project as discussed in table 4.5. The computational tests are divided into two subtests. In a first simulation run, the accuracy of the forecasting techniques is analyzed under the assumption that the project will end sooner than expected (RD < PD). In a second simulation run, it is assumed that the project is behind schedule, i.e. RD > PD. Consequently, the simulations only focus on the scenarios that measure normal project behavior and exclude the misleading scenarios of the previous section.

Summary results have been displayed in figure 4.6 for early (upper graph) and late (bottom graph) projects for the early, middle and late stages defined as the [0% - 30%], [30% - 70%] and [70% - 100%] percentage completed and for the PV2, ED2 and ES2 methods.

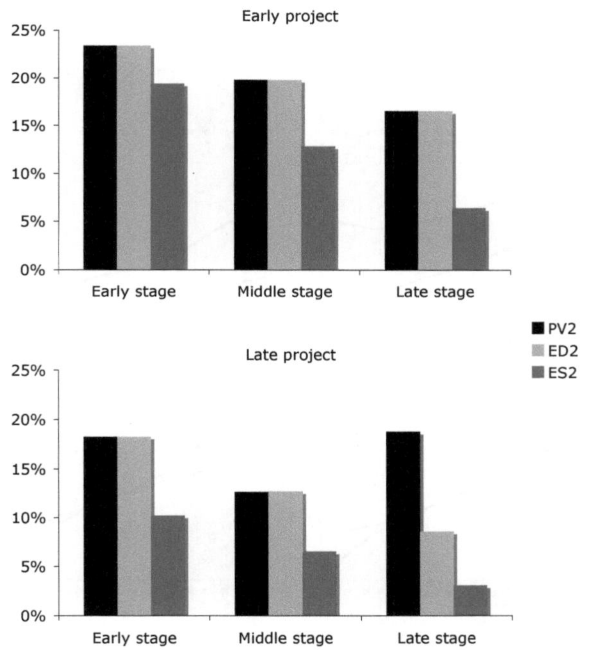

Fig. 4.6 The MAPE for early and late projects along the project completion stage

The results from the figure show that the earned schedule method outperforms, on average, the other forecasting methods. This is in line with the previously found results of section 4.2.2 where it has been concluded that the earned schedule method performs best under "normal" circumstances. The results also illustrate the quirky behavior of the SPI indicator (used in the planned value and earned duration methods) at the late stage of the project. Indeed, the late stage forecast accuracy is much better for the ES method compared to the PV and ED methods. The SPI(t) indicator of the earned schedule method is developed to overcome this quirky behavior, leading to an improved forecast accuracy at the end of the project. In the remainder of this section, more detailed results are discussed for both early and late projects based on the graphs displayed in figure 4.7. This figure displays the forecast accuracy (MAPE) of all methods with a performance factor equal to 1 (top graphs), SPI or SPI(t) (middle graphs) and SCI or SCI(t) (bottom graphs) along the life of the project, from the early to the late stages. Since it is hard to find a scale to fit all of the data while keeping visibility, the y-axes of the graphs might have different scales. The results can be summarized as follows:

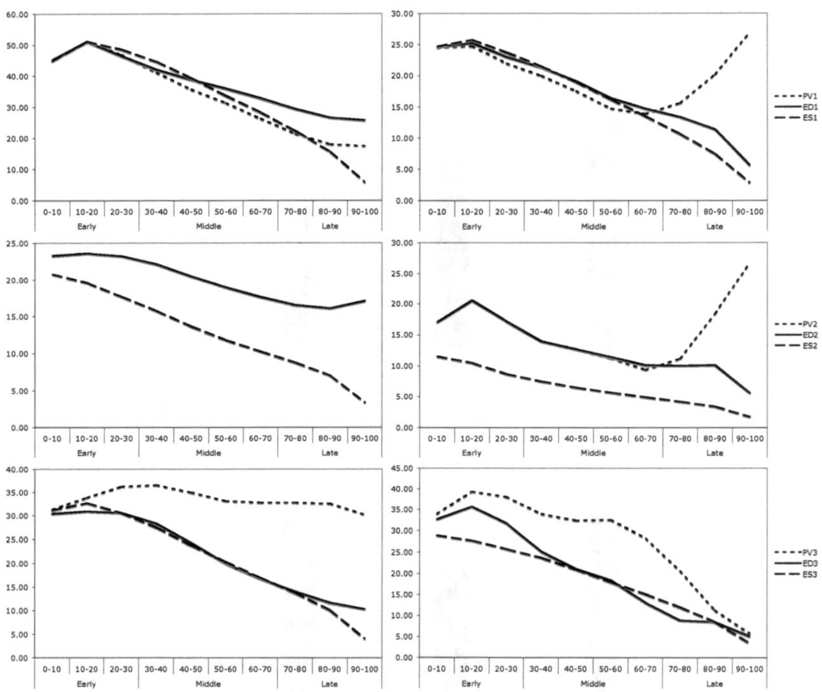

Fig. 4.7 The MAPE for early (left) and late (right) projects along the project completion stage

- All graphs clearly show that, obviously, the average forecast accuracy improves along the life of the project. The accuracy is rather poor in the early stages and gradually improves as the percentage completed goes up.
- The performance of all methods is comparable in the early and middle stages for the PV1, ED1 and ES1 methods.
- The earned duration method performs reasonably well when the performance factor is equal to one (ED1) or SCI (ED3).

- The graphs confirm the previously found results that the ES2 method outperforms all other methods, regardless of the stage of the project. Note that the scale of the y-axis (i.e. the forecast accuracy) is not equal for all graphs and shows the lowest percentages for the ES2 method.
- In the late stages of the project life cycle, all versions of the earned schedule method (i.e. ES1, ES2 and ES3) perform significantly better than the other methods, resulting in low absolute percentage errors. The planned value performs rather poor in the late stages and clearly shows a deteriorating trend towards the end of the project.

4.2.4 The influence of the network structure on the forecast accuracy

This section reports results for the relation between the forecast accuracy of all methods and the topological structure of the underlying project network. The motivation of this study lies in the conclusions of numerous papers where researchers have shown that the topological structure of a network has a clear influence on, for example, the constructed schedule (Patterson, 1976), the risk for delays (Tavares et al, 1999), the criticality of a network (Tavares et al, 2004) or the computational effort an algorithm needs to schedule a project (e.g. Elmaghraby and Herroelen (1980), amongst many others). The intention of this section is to test whether the forecast accuracy of the predictive duration methods depends on the four topological indicators presented earlier.

The serial/parallel indicator SP

The closeness of a project network to a serial or parallel network is measured by the serial/parallel indicator, SP $\in [0, 1]$. Obviously, the SP indicator is directly linked with the number of critical activities in a network (the closer the SP to 1, the more potential critical activities in the network). Consequently, this indicator can easily serve as an aid to detect in which cases the project level approach of EVM suffers from the more detailed activity level approach of the traditional critical path method (more details will be provided in chapters 5 and 6). This is motivated by Jacob and Kane (2004) who argue that earned value metrics and the corresponding forecasting indicators cannot be used but on the level of an individual activity. Indeed, a delay in a non-critical activity might give a false warning signal to the project manager, and hence, wrong corrective actions can be taken. However, a project manager is usually interested in the status of the overall project and has no time to calculate every metric on the activity level for practical reasons. Consequently, the EVM performance measures are calculated on the project level, and not on the level of each individual activity. The possible bias of this approach (project level) compared to the ideal approach (on the activity level) is influenced by the structure of the network, and more precisely by the number of critical activities in the networks.

Figure 4.8 displays the MPE for the networks with varying values for the SP indicator, ranging from 0.1 (close to a parallel network) to 0.9 (close to a serial network), in steps of 0.1. These graphs are ranked in a similar way as the 9 scenarios of table 4.2 and read as follows: the three graphs in the first row display results for scenarios 1, 4 and 7, the three graphs in the second row display results for scenarios 2, 5 and 8, etc. MPE values larger (lower) than zero give an indication of an overestimation (underestimation) of the forecasting metrics for the actual duration. Since it is hard to find a scale to fit all of the data while keeping visibility, the y-axes of the graphs might have different scales.

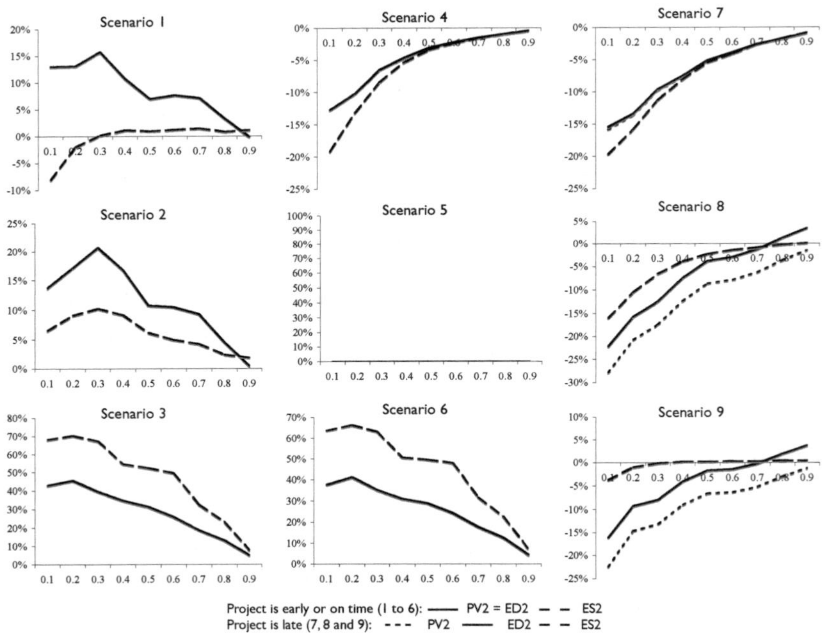

Fig. 4.8 The influence (MPE) of the serial or parallel networks for the 9 scenarios

The results of figure 4.8 can be summarized as follows. First, the tables reveal that the network structure clearly influences the forecast accuracy. Indeed, almost all graphs show an improving forecast performance (closer to zero) for all forecasting methods for increasing values of SP (i.e. more serial networks). The main reason is that the number of non-critical activities decreases for increasing SP values, and hence, the probability to make wrong conclusions decreases (delays in non-critical activities were the cause of misinterpretations as shown by Jacob and Kane (2004)). Second, most graphs reveal that the SPI or SPI(t) indicator is an important factor of the forecasting formulas which might give a reliable or unreliable warning signal. As an example, scenarios 3 and 6 are two misleading project scenarios finishing early or on time. However, the average SPI(t) value reports an expected project lateness which results in clear overestimations of the final project duration up to 70%. Scenarios 1 and 2 are similar scenarios that also finish early or on time. However, the average SPI(t) indicator correctly reports the excellent project performance, which results in maximal deviations of only 10%. A similar conclusion can be drawn for the earned schedule method for the scenarios 4, 7, 8 and 9 representing project finishing on time or late. As an example, scenario 7 finishes late, although the SPI(t) reports the opposite, which results in a low accuracy of the ES2 method and an average underestimation of the final project duration. Scenario 9, on the contrary, finishes late as correctly reported by the SPI(t) indicator, resulting in a rather accurate project duration forecast. Note that scenario 5 is an ideal scenario, with no deviations whatsoever, resulting in a 0% MPE value.

The influence of the three other network structure indicators is not always intuitively clear and often less outspoken than the SP indicator. In the remainder of this section, the main conclusions for the AD, LA and TF indicators are briefly summarized. The values for these indicators range from 0.2 to 0.8, in steps of 0.2. The results can be summarized as follows.

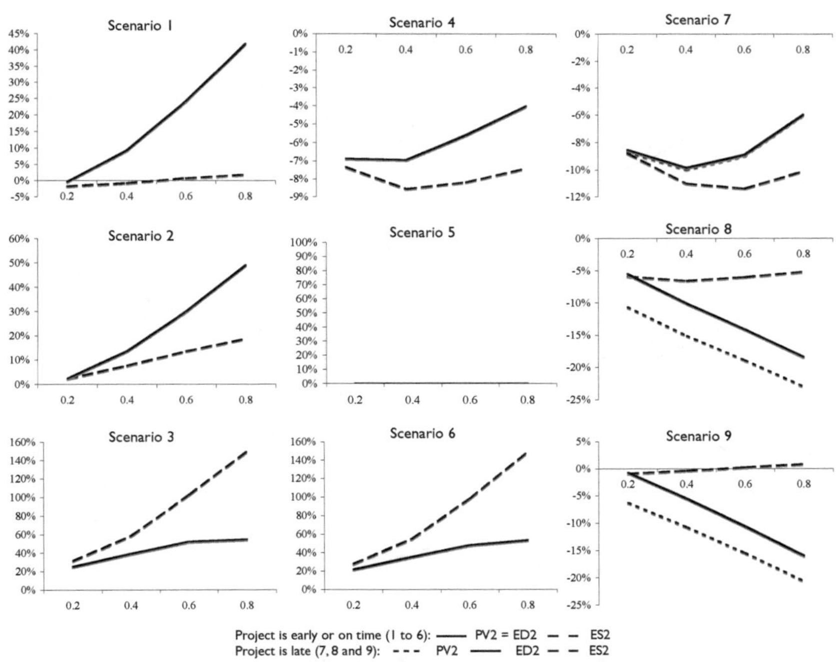

Fig. 4.9 The influence (MPE) of the AD indicator for the 9 scenarios

The influence of the activity distribution (AD)

The AD indicator shows a clear predictive pattern for the forecast accuracy of the three predictive methods (see figure 4.9 for the MPE values) showing more inaccurate forecasting results as the AD value goes up. This is a clear indication that the more the activities are spread out over the network (i.e. low AD values), the better, on the average, the forecast accuracy is. The ES method, however, is often stable and more reliable for the correct project scenarios (certainly for scenarios 1, 8 and 9), and its performance does not depend on the network structure measured by the AD indicator. Overall, the effect of the AD indicator is not so intuitively clear compared to the SP indicator.

The influence of the length of arcs (LA) and the topological float (TF)

The results of the LA and TF indicators reveal that these indicators have not always a significant effect on the performance of the forecasting metrics, as summarized in figures 4.10 and 4.11. The graphs of figure 4.10 display the effect of the LA indicator on the accuracy of the projects. One could expect an improving accuracy

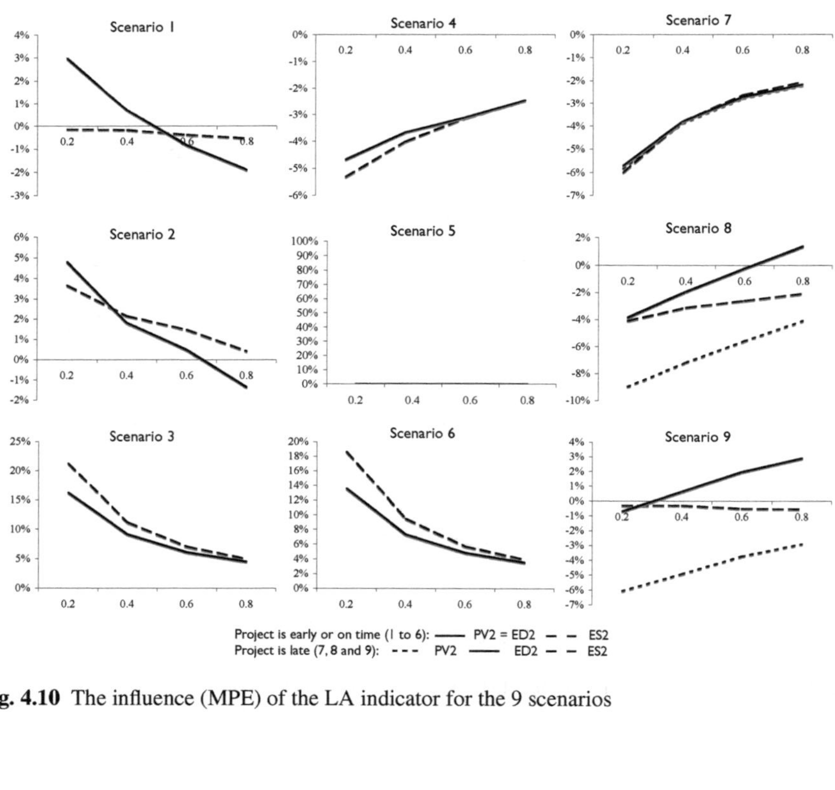

Fig. 4.10 The influence (MPE) of the LA indicator for the 9 scenarios

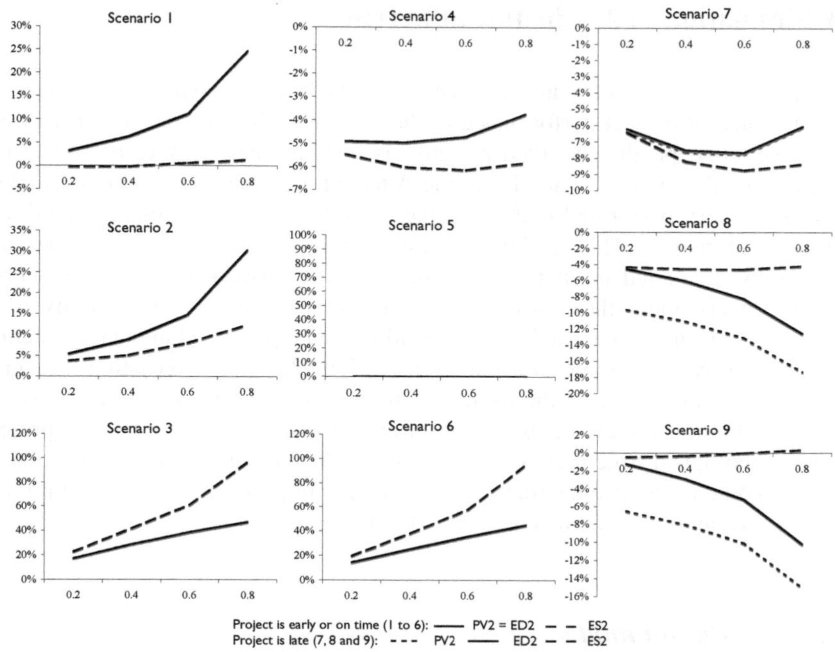

Fig. 4.11 The influence (MPE) of the TF indicator for the 9 scenarios

trend as the LA value increases. Indeed, higher LA values mean shorter precedence relations and consequently, it is more likely that the activity float is smaller. Hence, the occurrence of the error mentioned by Jacob and Kane (2004) is more unlikely, resulting in a better forecast accuracy. Although some graphs confirm this conjecture, it is not clear what the real effect of the LA indicator is.

For some metrics, the forecast accuracy displayed in figure 4.11 shows a deteriorating trend for increasing TF indicator values. The intuitive meaning of the TF indicator is that higher values for TF result in less dense networks since many activities can be shifted in the network. Consequently, an intuitive logic of this deteriorating effect could be that it is more likely that the increasing float of each activity with increasing values of TF could lead to more frequently occurring errors as mentioned by Jacob and Kane (2004) when calculating the EV metrics on the project level.

It should be noted that many effects of the LA and TF indicators are statistically insignificant and the forecast accuracy is often rather insensitive towards these indicator values.

4.3 Simulation 2: A schedule adherence study

In chapter 2, a recently introduced concept in earned value management to measure the adherence of project performance to the baseline schedule has been reviewed. This novel concept, the so-called p-factor, is a simple and easy extension to the earned schedule approach and allows the detection of project impediments and/or portions of work performed under risk based on the calculation of the traditional earned value metrics. This section elaborates on this schedule adherence measure and presents a detailed simulation study in which the network structure, the degree of activity overlapping, the earned value accrue and the variation in the activity duration will be carefully controlled. The simulation model presented in the next section aims at measuring the forecast accuracy of various duration forecasting measures with and without the use of the p-factor. The aim of this second simulation study is twofold. First, the study maps the evolution of the p-factor along various project network topological structures (see section 4.3.2). Second, the study tests the ability of the schedule adherence measure (the p-factor) to improve the accuracy of project duration forecasts (see sections 4.3.3 and 4.3.4).

4.3.1 Simulation model

In section 2.1 of chapter 2, three types of possible reasons that cause lack of schedule adherence have been briefly discussed, i.e. activity overlapping, difference between the PV and EV accrue and unexpected changes in the original activity time estimations. These causes will be reviewed here and implemented in the simulation model of this chapter.

Activity overlaps are simulated using a maximal percentage of the original activity duration estimation that overlaps with one or more predecessors, randomly varying between 0% and 50%. It is assumed that an activity can never finish before the finish of all its feeding (predecessor) activities, due to the inherent nature of the precedence relation between these activities. If unexpected delays in upstream activities endanger this assumption, activity pre-emption (splitting) is added to the downstream activity (see figure 4.12).

Fig. 4.12 Unexpected delay in activity 5 causes activity pre-emption in activity 9

The **EV and PV accrue** is simulated under three different settings. The planned value of each activity is set as a linear accrue of the unit cost per time unit. The earned value accrue can follow a similar linear behavior (no deviation), or follows a convex (slower initial performance followed by a catch-up) or concave (faster initial performance) accrue. To that purpose, the algorithm simulates an activity EV accrue within a restriction of the maximal deviation over the complete activity duration between its EV and its linear PV. In the test design, this restriction has been fixed at 20%, as displayed in figure 4.13 for activity 2 of figure 2.1.

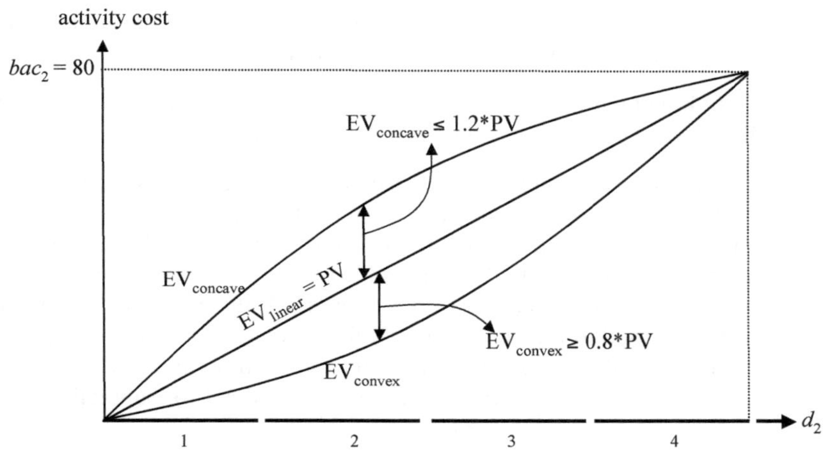

Fig. 4.13 Linear, convex and concave EV accrue

As mentioned previously, the random variation in activity durations is simulated by means of triangular distributions tailed to the right (activity delays) or to the left (activities ahead of schedule). The input parameters of triangular distributions have been carefully selected with a lower limit *a*, a mode *c* and an upper limit *b* resulting in 6 test case scenarios. The first three scenarios assume no activity overlapping and a linear EV/PV accrue and can be described as follows (d_i is used to denote the original duration estimation for each activity *i* of the project):

- Scenario 1: Random activity time/cost deviation: $(a,c,b) = (0.5*d_i, d_i, 1.5*d_i)$, which might result in project durations which are ahead, on or behind schedule.
- Scenario 2: Activities are, on average, ahead of schedule: $(a,c,b) = (0.1*d_i, d_i, 1.2*d_i)$ resulting in a total project duration ahead of schedule.
- Scenario 3: Activities have, on average, delays: $(a,c,b) = (0.8*d_i, d_i, 1.9*d_i)$ resulting in total project duration delays.

The next two scenarios assume no variation in activity time and cost estimations and hence, assume that all activity estimations are 100% correct (similar to scenario 5 of the first simulation study of this chapter). However, the presence of activity overlapping or EV/PV accrue deviations still might result in schedule changes and lack of schedule adherence. Consequently, these scenarios measure the influence of activity overlapping (scenario 4) and non-linear EV/PV accrue (scenario 5) on the forecast accuracy.

- Scenario 4: No time/cost deviations, $(a,c,b) = (d_i, d_i, d_i)$, maximal 50% activity overlapping and linear EV/PV = 1 accrue.
- Scenario 5: No time/cost deviations, $(a,c,b) = (d_i, d_i, d_i)$, no activity overlapping and a $0.8 \leq$ EV/PV ≤ 1.2 accrue.

Scenario 6 is a scenario where both activity duration deviations (similar to scenario 1) as well as activity overlapping and non-linear EV/PV accrue are taken into account. The scenarios can be summarized as follows:

- Scenario 6: Random time/cost deviations, $(a,c,b) = (0.5 * d_i, d_i, 1.5 * d_i)$, maximal 50% activity overlapping and a $0.8 \leq$ EV/PV ≤ 1.2 accrue.

In the next sections, the computational results for the second simulation study are only reported for the $\text{EAC(t)}_{\text{ES2}}$[4] measure. In subsection 4.3.2, the relation between the p-factor and the topological network structure is measured. In section 4.3.3, the simulation results are used to detect possible relations between the p-factor and the accuracy of the duration forecasts. In section 4.3.4, the effective earned value (EV(e)) method is used to predict a project's final duration. All runs for the three subsections are ran on the 6 scenarios described earlier and summarized in table 4.6. Rework has only been simulated in subsection 4.3.4, and not for subsections 4.3.2 and 4.3.3.

Table 4.6 A summary of the 6 simulation scenarios

Scenario	(a,c,b) as % of activity duration	Maximum % overlap	Minimum/maximum EV/PV accrue
1	(50%,100%,150%)	0%	1
2	(10%,100%,120%)	0%	1
3	(80%,100%,190%)	0%	1
4	(100%,100%,100%)	50%	1
5	(100%,100%,100%)	0%	0.8-1.2
6	(50%,100%,150%)	50%	0.8-1.2

4.3.2 The p-factor evolution and topological structure

In the current section, no rework is simulated. Instead, the evolution of the p-factor is measured for a wide variety of project networks to investigate the ability of the p-factor to measure schedule adherence. All figures in this section display the average

[4] Different robustness checks have been run by testing all other duration forecasting methods since it has been shown in the first simulation study that this method outperforms, on average, all other methods. No relevant deviations have been found.

p-factor (y-axis) as a function of the percentage completed (x-axis), measured as EV/BAC.

Figure 4.14 displays the average p-factor for the 6 scenarios (note that the y-axes of the graphs might have different scales). All graphs show the influence of the percentage critical/non-critical activities on the schedule adherence as measured by different SP $\in [0, 1]$ values (tests results are shown for a subset of the project network set, limited to project networks with values for the SP indicator equal to 0.2, 0.5 and 0.8). Project networks with low SP values have many activities in parallel and hence a lot of non-critical activities, while increasing SP values indicate a more serial project network and an increasing number of critical activities. An increased number of critical activities results in a higher p-factor indicating a better baseline schedule adherence. This observation is completely in line with Jacob and Kane (2004) who have argued that earned value analyses on the project level might lead to false interpretations due to neutralizing effects of critical versus non-critical activities. A higher percentage of critical activities obviously leads to a lower probability to make these errors, which results in a better forecast accuracy (see section 4.2.2). Figure 4.14 reveals that higher SP values also lead to a better schedule adherence. Indeed, since the schedule adherence of the p-factor is calculated relative to the ES metric, relative errors due to low SP values have an effect on the ES line and hence on the p-factor calculation. The more serial a project network is, the less degrees of freedom the project resources have to shift from activities suffering from constraints and/or impediments to the other activities in progress, resulting in smaller portions of work performed under risk.

The graphs also reveal that projects running ahead of schedule (scenario 2) have relatively lower p-factor values than projects with delays (scenario 3). This indicates that projects running ahead often have more work performed under risk relative to projects running behind schedule. This observation illustrates that the p-factor hardly says anything about good or bad project performance (one can hardly conclude that the low(er) p-factor for a project running ahead of schedule is due to bad project performance) but instead it simply measures the schedule adherence relative to the current project performance and allows the revelation of project impediments/constraints and possible rework. Note that all graphs show that, obviously, the p-factor grows as the percentage completed increases and ends at 1 at the project finish.

The figure also shows the p-factor evolution for projects with activity overlapping (scenarios 4 and 6) and with a convex and/or concave EV/PV accrue (scenarios 5 and 6). The graphs indicate that activity overlaps to up to 50% of the duration of the predecessors lead to p-factor values lower than one, indicating a lack of schedule adherence even when the real durations equal the original baseline activity durations. Intuitively, since no rework is simulated, not even for very large activity overlaps, the project execution deviates from the original baseline schedule. The effect of the difference between PV and EV accrue on the p-factor is less outspoken (scenario 5). Note that the graph for scenario 4 shows that, unlike any other graph, networks with a low SP value have a higher p-factor than networks with higher SP values for a percentage completion of 25%. Intuitively, low SP value networks have many

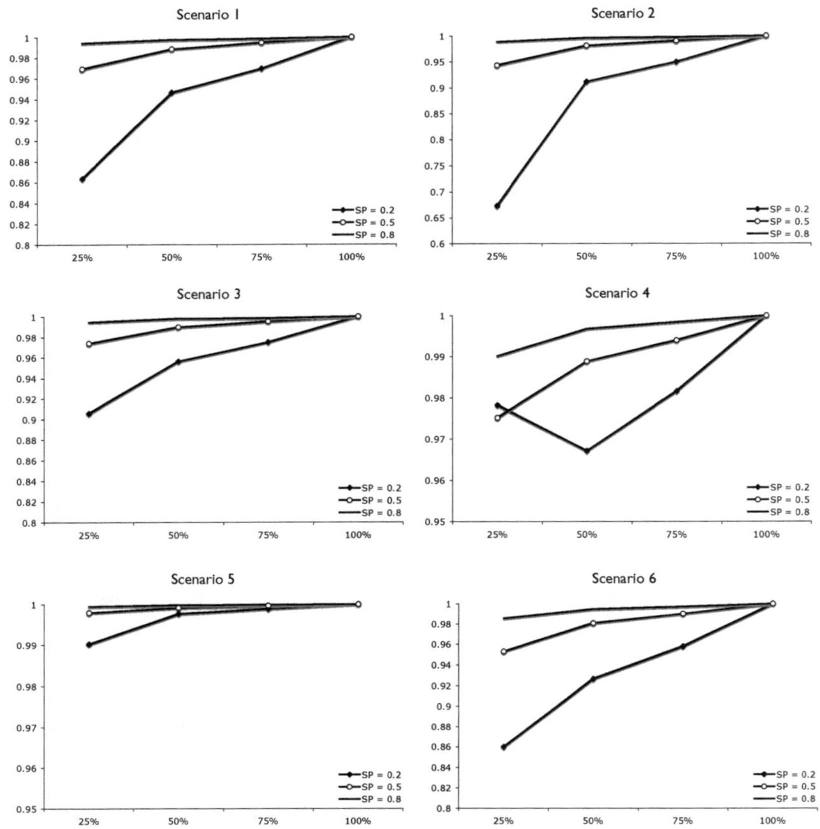

Fig. 4.14 p-factor evolution under 6 scenarios as a function of PC for three topological network structures (no simulated rework)

activities in parallel which leaves less room to overlaps with predecessor activities than more serial networks.

4.3.3 The p-factor and the duration forecasting accuracy

Since the p-factor can be considered as a measure for risk as a result of the lack of schedule adherence, the investigation of the relation between the average p-factor and the average forecast accuracy measured along the life of the project is an obvious step. Figure 4.15 displays the relation between the p-factor and the forecast accuracy measured by the MAPE for the 6 simulated scenarios. Similar to the previous section, no rework has been simulated. Since it is hard to find a scale to fit all of the data while keeping visibility, the y-axes of the graphs might have different scales.

The scatterplots of figure 4.15 all display a negative relation between the average p-factor and the forecast accuracy. All graphs show that lower p-factor values, denoting a certain lack of schedule adherence, often result in less accurate forecasts. Hence, the p-factor, which can be dynamically measured during a review of the project (i.e. project tracking) based on the traditional EVM metrics, can be considered as a warning signal of the duration forecast accuracy which can be seen as the most important result of the project tracking phase.

Consequently, while the SPI and/or SPI(t) metrics measure and indicate good (≥ 1) or bad (< 1) project performance since they compare current project execution with the original baseline schedule and can be used to predict a project's final duration, the p-factor value gives additional information since it measures the performance of the project relative to the earned schedule metric ES as a way to indicate schedule adherence. This factor can be used to validate the duration forecasts of the SPI or SPI(t) indicators, with lower p-factor values leading to less reliable predictions.

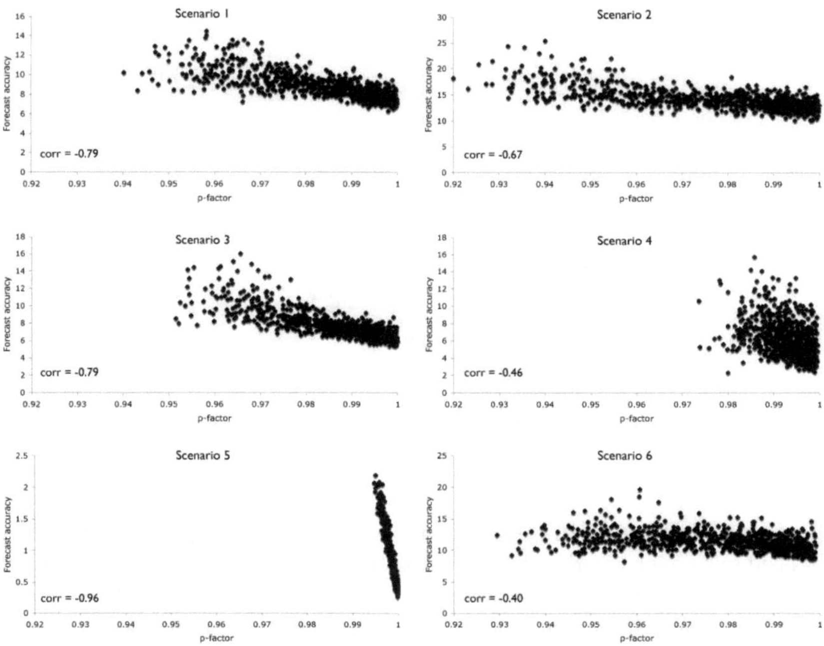

Fig. 4.15 The relation between the p-factor and the forecast accuracy for the 6 scenarios

4.3.4 The effective forecasting accuracy

This section reports results for the simulation study under the 6 different scenarios to test the duration forecast accuracy under the presence of rework with and without the use of the p-factor correction. Figure 4.16 shows the average forecasting accuracy with (MPE(e)) and without (MPE) the R% correction as used in the effective earned value. The figure is split into a left and right part to denote the forecast accuracy when underestimating or overestimating (note that for overestimations, RD < EAC(t) and MPE > 0 hold and vice versa for underestimations). Next to each bar, the relative percentage of forecasts with an under- or overestimation is displayed. Note that the rework correction R% has been set to 0.5 and the rework has been simulated in accordance with the R% value.

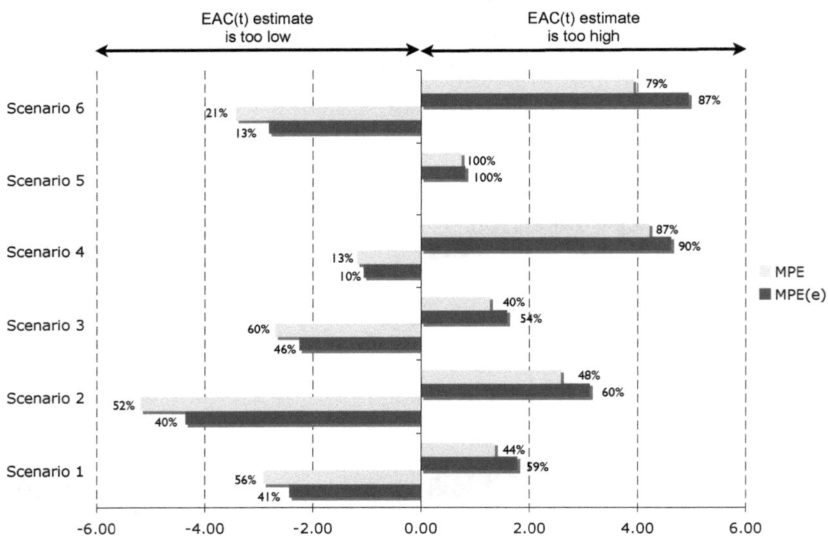

Fig. 4.16 The forecast accuracy (MPE) and the effective forecast accuracy (MPE(e))

The results in figure 4.16 can be summarized as follows. The MPE and MPE(e) rows need to be interpreted with care. Since each cell contains a different amount of observations (displayed as the number left or right to the bars), relative comparison between the 6 scenarios is hard to interpret. However, the graph clearly indicates the effect of the p-factor correction on the duration forecast accuracy as follows: it shows that the relevance of the p-factor to improve the duration forecast accuracy boils down to a correction factor to decrease the EV to an effective EV(e) leading to duration forecasting corrections. This forecasting correction means that optimistic forecasts (i.e. forecasts that predict a lower project duration than reality, displayed to the left of the y-axis) are corrected leading to, on average, an improved and less-optimistic forecast accuracy. However, pessimistic forecasts (bars with positive MPE or MPE(e) values) are unnecessarily corrected which even worsens the already pessimistic forecast. This observation is also shown in the decreased (increased) percentage of underestimations (overestimations) when introducing the R% correction factor. Consequently, although the p-factor is an easy and powerful tool to reveal project impediments and portions of work performed under risk, its ability to improve duration forecast accuracy is limited and should be taken with care.

4.4 Conclusion

This chapter has presented two simulation studies to test the ability and/or accuracy of the various methods presented in the previous chapters. The methodology used is Monte-Carlo simulation and results are obtained by running tests on an Apple MacBook with a Dual Core processor generating 100 simulation runs per setting. A large set of fictitious project networks has been used to generalize the computational results.

The first simulation study has tested the forecast accuracy of three different project duration forecasting methods, the planned value method, the earned duration method and the earned schedule method, based on extensive simulations on a large set of generated networks. First, the topological structure of the generated networks is carefully controlled based on an existing and reliable network generator, in order to make the link between the project network and the forecast accuracy. Second, the network structure is split up into critical and non-critical activities to measure their influence on the forecast accuracy. Last, the behavior of the SPI(t) indicator is varied to act as a correct or false warning signal, and hence, the type of uncertainty (risk) has been carefully controlled, resulting in 9 different test scenarios. The results reveal that the earned schedule method outperforms, on the average, all other forecasting methods. The closeness of a network to a serial or parallel network directly influences the activity slack and has an impact on the accuracy of the forecasts. This research is highly relevant to both academicians and practitioners. From an academic point of view, it is interesting to measure the influence of the network structure on the behavior of both scheduling and monitoring tools, and hence, this research serves as a call to researchers to focus their further research attention towards specific problems or project instances. Indeed, rather than developing tools and techniques for general problems that have, on average, an excellent behavior, one can better focus on the development of tools and methods for a set of specific problem instances, which belongs to a certain class or set of network structures. This research area is closely related to the phase transition research attention that has been described in many research papers (see e.g. Herroelen and De Reyck (1999) who describe the concept of phase transitions from a project scheduling point of view). The research to evaluate the forecast accuracy of project duration forecasting methods is completely void. Hence, this study can be used by practitioners who use these metrics and principles on a daily basis, but are unaware of the merits and the pitfalls of each individual method. Moreover, this research chapter also provides a framework for future research purposes and hence, can be used as a guide for EVM researchers.

In a second simulation study, a recently introduced concept in earned value management to measure the adherence of project performance to the baseline schedule is reviewed and tested through new simulation runs. This novel concept, the so-called p-factor, is a simple and easy extension to the earned schedule approach and allows the detection of project impediments and/or portions of work performed under risk based on the calculation of the traditional earned value metrics. A detailed simulation study has been run while carefully controlling the network structure, the degree of activity overlapping, the earned value accrue and the variation in the activity duration. All results measure the forecast accuracy of various duration forecasting measures with and without the use of the p-factor. This research is highly relevant to both academicians and practitioners and comes towards the need of a more profound study of newly developed concepts for which preliminary results and conclusions are often based on arbitrary cases or loose statements without a detailed theoretical knowledge and/or empirical observations based on runs on extensive sets of data. The relevance of the p-factor concept mostly lies in the easy and simple detection of portions of work that show either constraints and impediments or performance under risk, relative to the baseline schedule. The results have also shown that the simple dynamic calculation of the p-factor (based on the traditional EV metrics) might help to predict the accuracy of the forecasts along the life of the project. However, the use of this factor to improve forecast accuracy is limited to incorporating corrections for optimistic duration forecasts by adjusting the earned value to an effective earned value taking the possibility of rework into account. In doing so, the forecast accuracy of optimistic scenarios will be corrected to more realistic estimations while pessimistic forecasts will suffer under this correction mechanism.

Chapter 5
Time Sensitivity

The interest in activity sensitivity from both the academics and the practitioners lies in the need to focus a project manager's attention on those activities that influence the performance of the project. When management has a certain feeling of the relative sensitivity of the various parts (activities) on the project objective, a better management's focus and a more accurate response during project tracking should positively contribute to the overall performance of the project.

In this chapter, a third simulation study is performed to measure the ability of four basic sensitivity metrics to dynamically improve the time performance during project execution. Activity sensitivity information is used to guide the corrective action decision making process to improve a project's time performance, while varying the degree of management's attention. A large amount of simulation runs are performed on a large set of fictitious project networks generated under a controlled design.

5.1 Introduction

Since the introduction of the well-known PERT in the late 1950's in project scheduling, the research on measuring a project's sensitivity has been increasingly received attention from both practitioners and academics. Motivated by the common knowledge that the traditional critical path analysis gives an optimistic project duration estimate (see e.g. Klingel (1966), Schonberger (1981), Gutierrez and Kouvelis (1991) and many others), measuring the project sensitivity and the ability to forecast the final duration during its execution have become key parameters for project managers. Despite the many often diverse research outputs, shortcomings are mentioned from different research angles and a lot of confusions on advantages and/or disadvantages have been mentioned. However, only partial answers on the shortcomings have been reported in literature.

The aim and contribution of this chapter is twofold. First, the chapter briefly reviews basic as well as more advanced sensitivity measures used throughout the

M. Vanhoucke, *Measuring Time*, International Series in Operations Research & Management Science 136, DOI: 10.1007/978-1-4419-1014-1_5,

literature. Second, the relation between the project duration sensitivity and the ability and accuracy of forecasting a project's final duration is investigated in detail. A third simulation study is performed to measure the usefulness of sensitivity measures during project tracking and to provide general guidelines to project managers where and when activity-based sensitivity measures are useful as a dynamic tool to support the corrective action decision making process during project tracking. The results are obtained by simulation runs on the set of artificial projects generated in the previous chapter.

The outline of this chapter is as follows. Section 5.2 reviews the most important research efforts on sensitivity measures in project scheduling and presents illustrative examples. In section 5.3, a simulation study is presented in which the relation between sensitivity measures and the link to EVM forecast accuracy is tested in detail. Section 5.4 gives overall conclusions.

5.2 Literature overview

This section provides a general summary overview of the research on activity and project sensitivity in project scheduling. In section 5.2.1, four sensitivity measures used in the study of this chapter are discussed in detail. Sections 5.2.2 and 5.2.3 present example calculations and highlight advantages and disadvantages. Section 5.2.4 briefly repeats the earned value based forecasting results obtained from the previous chapter that will be used in the simulation study presented here.

5.2.1 Activity-based sensitivity measures

The literature on project sensitivity measures is wide and diverse and focuses on the measurement of the relative activity sensitivity in relation to the project duration. Typically, many papers and handbooks mention the idea of using Monte-Carlo simulations as the most accessible technique to estimate a project's completion time distribution. These research papers present often simple metrics to measure a project's sensitivity under various settings. Williams (1992) reviews three important sensitivity measures to measure the criticality and/or sensitivity of project activities. Elmaghraby (2000) critically reviews these sensitivity measures and extends the domain to more fundamental sensitivity measures. In section 5.2.3, the more advanced studies discussed by Elmaghraby (2000) are briefly reviewed, but will not be further used in this book. Motivated by the heavy computational burden of simulation techniques, various researchers have published analytical methods and/or approximation methods as a worthy alternative. An overview can be found in the study of Yao and Chu (2007) and will not be discussed in the current chapter.

In this section, the three activity based sensitivity measures discussed in Williams (1992) are briefly reviewed. A fourth sensitivity measure, published in PMBOK

(2004) is added in the simulation experiment of this chapter. The following notation will be used throughout this chapter:

nrs	Number of Monte-Carlo simulation runs (index k)
d_i	Duration of activity i (superscript k will be used to refer to the d_i of simulation run k)
tf_i	Total float of activity i (superscript k will be used to refer to the tf_i of simulation run k)
RD	Real Duration of the project (superscript k will be used to refer to the RD of simulation run k) → often referred to as the project makespan

Criticality Index CI:

The criticality index measures the probability that an activity lies on the critical path. It is a simple measure obtained by Monte-Carlo simulations, and is expressed as a percentage denoting the likelihood of being critical. The concept was introduced by Martin (1965) and further extended by various authors (see e.g. Van Slyke (1963), Dodin and Elmaghraby (1985)) and Fatemi Ghomi and Teimouri (2002), amongst others). The CI of activity i can be given as follows:

$$CI = P(tf_i = 0) = \frac{\sum_{k=1}^{nrs} \begin{cases} 1 & \text{if } tf_i^k = 0 \\ 0 & \text{otherwise} \end{cases}}{nrs} \tag{5.1}$$

Although the criticality index has been used throughout various studies and implemented in many software tools, the CI often fails in adequately measuring the project risk. The main drawback of the CI is that its focus is restricted to measuring probability, which does not necessarily mean that high CI activities have a high impact on the total project duration (e.g. think of a very low duration of an activity always lying on the critical path, but with a low impact on the total project duration due to its negligible duration).

Significance Index SI (Williams, 1992):

In order to better reflect the relative importance between project activities, the sensitivity index of activity i has been formulated as follows:

$$SI = E\left(\frac{d_i}{d_i + tf_i} * \frac{RD}{E(RD)}\right) \tag{5.2}$$

with $E(x)$ used to denote the expected value of x. The SI has been defined as a partial answer to the criticism on the CI. Rather than expressing an activity's criticality by the probability concept, the SI aims at exposing the significance of individual activities on the total project duration. In some examples, the SI seems to provide more acceptable information on the relative importance of activities. Despite this, there are still examples where counter-intuitive results are reported (see section 5.2.3).

Cruciality Index CRI (Williams, 1992):

A third measure to indicate the duration sensitivity of individual activities on the total project duration is given by the correlation between the activity duration and the total project duration, as follows:

$$CRI = |corr(d_i, RD)| \tag{5.3}$$

This measure reflects the relative importance of an activity in a more intuitive way and calculates the portion of total project duration uncertainty that can be explained by the uncertainty of an activity. This correlation measure can be easily calculated by using the Pearson's product-moment. However, this correlation metric is a measure of the degree of linear relationship between two variables. However, the relation between an activity duration and the total project duration often follows a non-linear relation. Therefore, Cho and Yum (1997) propose to use non-linear correlation measures such as the Spearman rank correlation coefficient or Kendall's tau measure. These three correlation measures can be calculated as follows:

1. *Pearson's product-moment* of activity i can be calculated as follows:

$$r = \frac{\sum_{k=1}^{nrs}(d_i^k - \bar{d}_i)(RD^k - \overline{RD})}{nrs * \sigma_{d_i} * \sigma_{RD}} \tag{5.4}$$

   with $\bar{x}$ the average and σ_x the standard deviation of variable x[1].

2. The *Spearman's rank correlation* assumes that the values for the variables are converted to ranks, followed by the calculation of the difference between the ranks of each observation on the two variables. The measure is equal to:

$$\rho = 1 - \frac{6\sum_{k=1}^{nrs}\delta_k^2}{nrs(nrs^2 - 1)} \tag{5.5}$$

   where δ_k is the difference between the ranking values of d_i and RD during simulation run k.
3. *Kendall's tau rank correlation* index measures the degree of correspondence between two rankings as follows:

[1] Note that this formula assumes that the measure is calculated using population standard deviations which explains the nrs in the denominator. When using sample standard deviations, the formula should be divided by nrs - 1 instead.

$$\tau = \frac{4P}{nrs(nrs-1)} - 1 \tag{5.6}$$

where P is used to represent the number of concordant pairs[2] of the d_i and RD variables.

Schedule Sensitivity Index SSI (PMBOK, 2004):

The Project Management Body Of Knowledge (PMBOK) mentions quantitative risk analysis as one of many risk assessment methods, and proposes to combine the activity duration and project duration standard deviations (σ_{d_i} and σ_{RD}) with the criticality index. In this chapter, it will be referred to as the schedule sensitivity index and it is equal to:

$$SSI = \frac{\sigma_{d_i} * CI}{\sigma_{RD}} \tag{5.7}$$

5.2.2 An illustrative example

Table 5.1 shows 5 fictitious simulated scenarios for the example project network of figure 1.7. Each scenario is characterized by a set of activity durations and a total real project duration RD. Note that scenario 1 corresponds to the Gantt chart presented in figure 1.9.

Table 5.2 displays the values for all sensitivity measures and table 5.3 displays the intermediate calculations required to calculate the sensitivity measures.

Table 5.1 5 simulation scenarios to perform a schedule risk analysis

Scenario	**2**	**3**	**4**	**5**	**6**	**7**	**8**	**9**	**10**	**11**	**RD**
1	6	11	1	4	3	2	8	8	5	4	18
2	7	14	1	4	5	1	8	11	5	5	22
3	6	16	1	7	8	1	11	14	6	5	27
4	4	14	1	7	5	1	9	9	5	3	20
5	6	17	2	6	9	2	13	12	4	4	26
Average	5.8	14.4	1.2	5.6	6	1.4	9.8	10.8	5	4.2	22.6
StDev.	0.98	2.06	0.40	1.36	2.19	0.49	1.94	2.14	0.63	0.75	3.44

The sensitivity measures are calculated for illustrative purposes for activity 2 of the example network.

[2] Let (x_i, y_i) and (x_j, y_j) be a pair of (bivariate) observations. If x_j - x_i and y_j - y_i have the same sign, the pair is *concordant*, if they have opposite signs, the pair is *discordant*.

Table 5.2 The sensitivity measures for all activities obtained through a schedule risk analysis

	2	3	4	5	6	7	8	9	10	11
CI	0.80	0.00	0.20	0.80	0.20	0.20	0.20	0.80	0.00	0.00
SI	0.94	0.82	0.36	0.94	0.61	0.38	0.72	0.97	0.62	0.30
CRI (r)	0.27	0.93	0.49	0.52	0.96	0.14	0.83	0.97	0.09	0.50
CRI (ρ)	0.30	0.88	0.50	0.50	0.88	0.13	0.73	1.00	0.30	0.60
CRI (τ)	0.20	0.60	0.40	0.20	0.60	0.60	0.40	1.00	0.20	0.20
SSI	0.23	0.00	0.02	0.32	0.13	0.03	0.11	0.50	0.00	0.00

Criticality Index CI: The row with label "critical (yes/no)" displays for each scenario whether the activity is critical or not, and is used to calculate the criticality index. As an example, the CI for activity 2 is equal to CI = $\frac{4}{5}$ = 0.80.

Significance Index SI: The activity float (row "Total Float") is necessary to calculate the significance index as SI = $(\frac{6}{6+0} * \frac{18}{22.6} + \frac{7}{7+0} * \frac{22}{22.6} + \frac{6}{6+0} * \frac{27}{22.6} + \frac{4}{4+0} * \frac{20}{22.6} + \frac{6}{6+2} * \frac{26}{22.6})$ / 5 = 0.94.

Cruciality Index CRI: The cruciality index CRI can be calculated using the three formulas:.

- **CRI (r):** The CRI (r) measure is calculated as CRI (r) = $\frac{1}{5*0.98*3.44} * (6-5.8) * (18-22.6) + (7-5.8) * (22-22.6) + (6-5.8) * (27-22.6) + (4-5.8) * (20-22.6) + (6-5.8) * (26-22.6) = 0.27$.
- **CRI (ρ):** In order to avoid errors resulting from non-linearities, the CRI (ρ) and CRI (τ) require a transformation of the original data into a ranking. In case of tied ranks, the same rank is given to each of the equal values as the average of their positions in the ranking. As an example, placing the activity durations of activity 2 in increasing order for all scenarios results in the following scenario sequence 4 - $\overbrace{1-3-5}^{\textit{tie break}}$ - 2 corresponding to a ranking [2,5,3,1,4] for scenarios [1,2,3,4,5]. However, tie breaks occur for scenarios 1, 3 and 5. In this case, the average is taken of their ranking values as $\frac{2+3+4}{3} = 3$, resulting in the ranking [3,5,3,1,3] as shown in the rows with label "ranking (tie breaks)" of table 5.3. Consequently, the CRI (ρ) measure uses these rankings to calculate the δ values and is equal to CRI $(\rho) = 1 - 6 * \frac{(3-1)^2+(5-3)^2+(3-5)^2+(1-2)^2+(3-4)^2}{5*(5^2-1)} = 0.30$.
- **CRI (τ):** The CRI (τ) measure relies on the Kendall τ coefficient, in which the P value can be calculated rather easily by re-ordering the ranks in increasing order of the RD ranking values (cf. the rows with label "ranking (re-ordered)" of table 5.3). The P value is then calculated by counting for each scenario how many ranking values displayed below the current scenario are higher than the ranking for the current scenario. For example, scenario 1, only 1 ranking value (i.e. for scenario 2) below scenario 1 is higher than the current ranking value, and hence, the contribution to P is 1. For scenario 4, three ranking values displayed below this scenario have a higher ranking value, and hence, its contribution to P equals 3. Consequently, the P value for activity 2 is equal to 1 + 0 + 3 + 0 + 0 = 4 and CRI $(\tau) = |\frac{4*4}{5*(5-1)} - 1| = 0.20$.

Schedule Sensitivity Index SSI: The schedule sensitivity index can be calculated as $\mathrm{SSI} = \frac{0.98*0.80}{3.44} = 0.23$.

Table 5.3 Intermediate calculations for the sensitivity measures

	Scenario	2	3	4	5	6	7	8	9	10	11	RD
Critical (yes/no)	1	yes	no	no	yes	no	no	no	yes	no	no	-
	2	yes	no	no	yes	no	no	no	yes	no	no	-
	3	yes	no	no	yes	no	no	no	yes	no	no	-
	4	yes	no	no	yes	no	no	no	yes	no	no	-
	5	no	no	yes	no	yes	yes	yes	no	no	no	-
Total Float	1	0	2	4	0	4	4	4	0	2	8	-
	2	0	3	7	0	7	7	7	0	3	10	-
	3	0	5	6	0	6	6	6	0	5	12	-
	4	0	1	4	0	4	4	4	0	1	10	-
	5	2	5	0	2	0	0	0	2	5	9	-
Ranking (tie breaks)	1	3	1	2.5	1.5	1	4.5	1.5	1	3	2.5	1
	2	5	2.5	2.5	1.5	2.5	2	1.5	3	3	4.5	3
	3	3	4	2.5	4.5	4	2	4	5	5	4.5	5
	4	1	2.5	2.5	4.5	2.5	2	3	2	3	1	2
	5	3	5	5	3	5	4.5	5	4	1	2.5	4
Ranking (re-ordered)	1	3	1	2.5	1.5	1	4.5	1.5	1	3	2.5	1
	4	1	2.5	2.5	4.5	2.5	2	3	2	3	1	2
	2	5	2.5	2.5	1.5	2.5	2	1.5	3	3	4.5	3
	5	3	5	5	3	5	4.5	5	4	1	2.5	4
	3	3	4	2.5	4.5	4	2	4	5	5	4.5	5
	P	4	8	3	6	8	2	7	10	4	6	-

Figure 5.1 displays the sensitivity measures of table 5.2 graphically. Note that these graphs are only displayed for illustrative purposes, and hence, no general results can be drawn.

Figure 5.2 illustrates how the CRI(r) sensitivity information of project activities can be used and how an action threshold can be set as a minimal threshold value of the sensitivity measure. This action threshold defines the degree of control, which can vary between no control and full control, and is shown by the vertical dotted line on the figure. All activities with a CRI(r) value higher than or equal to this line are said to be highly sensitive activities which require attention during the tracking process and corrective actions in case of delays. In the example case of the figure, the action threshold has been set to 50% such that only the most sensitive activities 3, 5, 6, 8 and 9 with a CRI(r) value higher than 0.5 need to be considered during the tracking process. In the remainder of this chapter, this will be referred to as a % Control value equal to 50%.

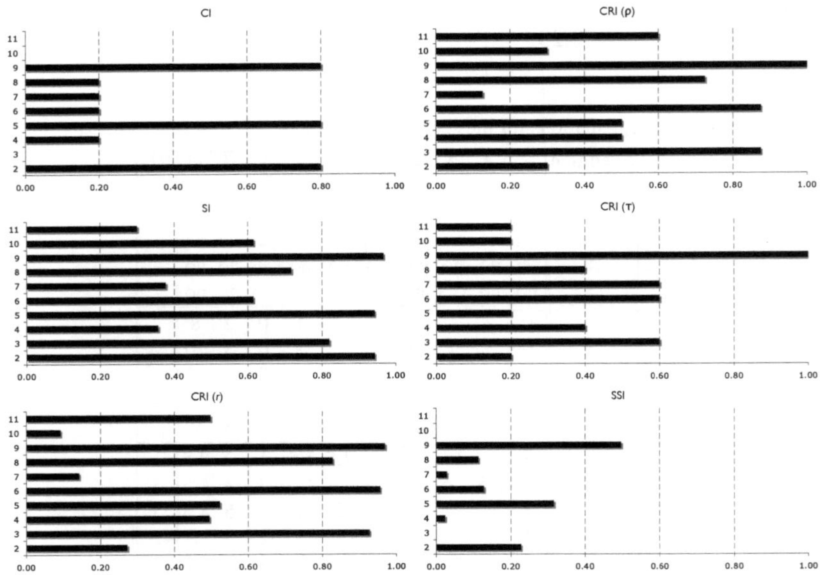

Fig. 5.1 Graphical representation of the sensitivity measures for the example network

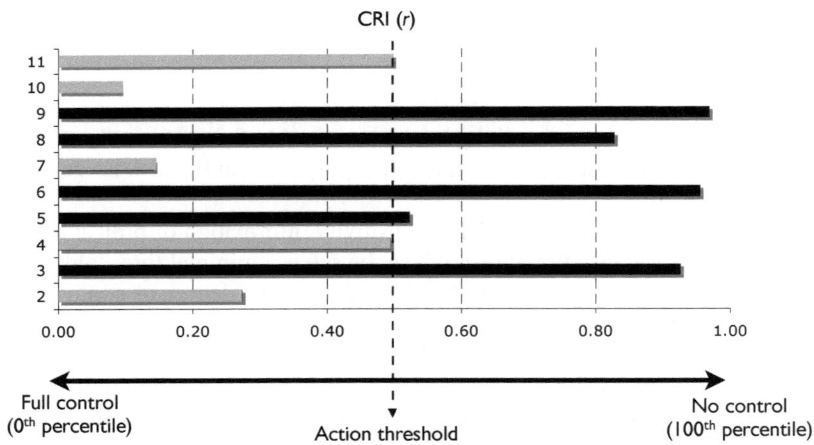

Fig. 5.2 Action threshold as a percentile between full and no control

5.2.3 A critical view on sensitivity measures

In this subsection, a critical eye is thrown to the four sensitivity measures discussed previously. Although the simulation study of this chapter will be restricted to the use of these four sensitivity measures, a short critical review is given here as a summary from various sources in literature. A detailed study of these sensitivity extensions is outside the scope of this chapter, and the reader is referred to the different sources mentioned in this section.

Williams (1992) shows illustrative examples for three sensitivity measures, CI, SI and CRI, and mentions weaknesses for each metric. For each sensitivity metric, anomalies can occur which might lead to counter-intuitive results. Numerous extensions have been presented in literature that (partly) give an answer on these shortcomings and/or anomalies. Tavares et al (2004) present a surrogate indicator of criticality by using a regression model in order to offer a better alternative to the poor performance of the criticality index in predicting the impact of an activity delay on the total project duration. Kuchta (2001) presents an alternative criticality index based on network information. However, no computational experiments have been performed to show the improvement of the new measure. In Elmaghraby (2000), a short overview is given on the advantages and disadvantages of the three sensitivity measures discussed in Williams (1992). He conjectures that a relative importance of project activities should be given by considering a combined version of these three sensitivity measures and reviews the more advanced studies that give partial answers on the mentioned shortcomings. More precisely, the paper reviews the research efforts related to the sensitivity of the mean and variance of a project's total duration due to changes in the mean and variance of individual activities. Cho and Yum (1997) propose an uncertainty importance measure to measure the effect of the variability in an activity duration on the variability of the overall project duration. Elmaghraby et al (1999) investigate the impact of changing the mean duration of an activity on the variability of the project duration. Finally, Gutierrez and Paul (2000) present an analytical treatment of the effect of activity variance on the expected project duration.

The use of the criticality index CI has been criticized throughout literature since it is based on probabilistic considerations which are very far from management's view on the project. Moreover, the metric only considers probabilities, while it is generally known that the risk of an activity depends on a combination of probability and impact. The latter is completely ignored in the CI value, as illustrated in figure 5.3. The figure shows a parallel project network (the non-numbered nodes are used to denote the start and end dummy activities) with the possible durations and the corresponding probabilities denoted above each node. Obviously, activity 1 is the most sensitive activity since it mainly is responsible for the total project risk. However, the CI of activity 1 is equal to 1%, which is much lower than the CI = 99% of activity 2.

Although the SI and CRI measures have been proposed to reflect the relative importance of an activity in a better way than the CI, they can, however, both produce counter-intuitive results as illustrated by means of the example network of

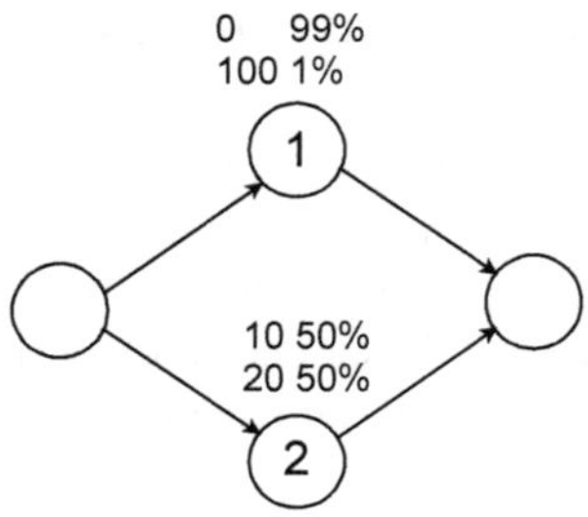

Fig. 5.3 A parallel two non-dummy activity example network (SP = 0) (Source: Williams (1992))

figure 5.4. Clearly, activity 1 has the largest impact on the project duration and $E(RD) = 115$. However, the SI values are equal for both activities and hence no distinction is made between the sensitivity of both activities. Indeed, the SI is equal to $100\% * \frac{100}{100} * \frac{115}{115} = 1$ for activity 1 and to $50\% * \frac{10}{10} * \frac{110}{115} + 50\% * \frac{20}{20} * \frac{120}{115} = 1$ for activity 2. Even worse, the CRI values show an opposite risk profile for both activities. The CRI measure shows only the effect on the risk of the total project and, consequently, if the duration of an activity is deterministic (or stochastic but with very low variance), then its CRI is zero (or close to zero) even if the activity is always on the critical path. The CRI value for activity 1 is equal to 0% (no variation) while it is equal to $\frac{(10-15)*(110-115)+(20-15)*(120-115)}{2*5*5} = 1$ for activity 2.

Fig. 5.4 A serial two non-dummy activity example network (SP = 1) (Source: Williams (1992))

5.2.4 Earned Value forecasting accuracy

The simulation study of section 4.2 measures and evaluates the accuracy of earned value management methods to predict the final duration of a project. Two main conclusions of this simulation study are crucial for this chapter and are briefly repeated hereunder:

- The earned schedule based schedule performance index SPI(t) is a rather stable indicator along the whole life of the project, and is more reliable for serial projects than for parallel projects.

- The earned schedule method is, on average, the most reliable EVM based project duration forecasting method (compared to the two other existing methods, the planned value method (Anbari, 2003) and the earned duration method (Jacob, 2003)) under normal project progress circumstances. The method predicts the final project duration as EAC(t) = AD + (PD - AD) / SPI(t).

Based on earned value information, the schedule performance index SPI(t) of a project gives overall information on the time performance of the project at the current time instance. This schedule performance index is set up as an early warning signal to detect problems and/or opportunities in an easy and efficient way and is measured at the cost account level, or even higher, rather than a simple replacement of the activity based critical path scheduling tools. Consequently, a project manager can decide to take corrective actions when a general project performance status drops below a critical threshold (e.g. $SPI(t) < 0.75$) to take corrective actions. If corrective actions are necessary, the project manager needs to drill down into lower WBS levels (up to the individual activity level) and take the appropriate corrective actions on those activities which are in trouble (especially those tasks which are on the critical path). In the remainder of this chapter, it is referred to this approach as the *project based tracking process* to express that general project based performance measures (SPI(t)) are used to trigger the activity based corrective action decision making process. Obviously, the reliability of the duration performance measures and forecasts on the project level are crucial during project tracking and affect the adequacy of the corrective action decision making process. Reliable forecasts allow the project manager to restrict the focus on simple project based sanity checks to trigger the often time-consuming critical path scheduling and tracking process.

However, when those project based EVM forecasts are unreliable, a more activity-based project tracking approach (i.e. at lower WBS levels than usually done for the EVM performance measures) is stringent which demands a stronger control on a larger subset of activities and a continuous critical path based tracking decision process. This approach will be further denoted as the *activity based tracking process* to express that individual activity information is required to effectively manage the project tracking and corrective action making decision process. In the next section, the ability to use the activity sensitivity information of section 5.2.1 as a supportive tool for an activity based project tracking approach is tested for a large set of projects under various simulation settings.

5.3 Simulation 3: An activity sensitivity study

This section presents results for the simulation runs to measure the adequacy of the four sensitivity measures of section 5.2.1 during the project tracking process. Section 5.3.1 presents the test design and the simulation scenarios. Section 5.3.2 presents details and results on the simulation study in which the sensitivity measures are dynamically used to improve the overall project performance.

5.3.1 Test design

The 4,100 project networks used in the simulation study have been discussed in the previous chapter, and their main characteristics will be briefly repeated here. The test set has been generated under a controlled test design by the activity-on-the-node network generator RanGen (Demeulemeester et al, 2003; Vanhoucke et al, 2008) and contains project networks with a varying topological structure. The serial/parallel (SP) indicator has been shown to be a relevant network topology indicator for the forecast accuracy of earned value based time prediction measures and plays a central role in the current simulation study.

Project execution is simulated for each project network using Monte-Carlo simulation runs under various scenarios. The input parameters a, b and c of a triangular distribution to simulate activity/cost variation have been carefully selected resulting in two test case scenarios described as follows (d_i^b is used to denote the original baseline duration estimate for each activity i of the project):

- Scenario 1: Random activity time deviations: $(a,c,b) = (0.5 * d_i^b, d_i^b, 1.5 * d_i^b)$, which might result in activity durations which are ahead, on or behind schedule. The projects finish, on average, on time.
- Scenario 2: Activity delays: $(a,c,b) = (0.8 * d_i^b, d_i^b, 1.9 * d_i^b)$ resulting in average total project duration delays.

The simulation runs of the next section test whether the activity-based sensitivity measures are a useful alternative to the poor project-based forecasting quality for projects with low SP values, and are an adequate tool for an activity based project tracking approach. All simulations are extended to a dynamic corrective action decision making process which uses the performance measurement status of the project as a trigger to take adequate and effective corrective actions.

5.3.2 Corrective actions

The design of the third simulation study is outlined in figure 5.5, split up into input parameters, simulation details and output measures. The first three steps (project data, run simulation and measure activity sensitivity) are obvious steps used in a traditional schedule risk analysis study (Hulett, 1996) and consist of 100 simulation runs for all project networks in order to obtain a value for all sensitivity measures as discussed in section 5.2.1.

The values for the sensitivity measures for all activities are used as action thresholds in the next simulation run (run simulation with corrective actions). More precisely, a subset of highly sensitive activities will be selected during project progress to measure their current time performance from the moment an action threshold has exceeded. The action threshold is set between the minimal sensitivity value and maximal sensitivity value (e.g. the average sensitivity) of all activities obtained from the first simulation run.

In this chapter, results are reported for action thresholds which are simple functions of the activity sensitivity measures. As an example, an action threshold can be set as the average or median sensitivity value of all project activities, such that approximately 50% of the activities are considered to be highly sensitive activities subject to the project activity tracking process. Obviously, other action threshold values can be set to increase (an action threshold closer to the minimum sensitivity value) or decrease (an action threshold closer to the maximum sensitivity value) the number of activities subject to the tracking process. This level of the action threshold defines the effort a project manager puts in the project tracking phase, as will be measured by the % Control output measure discussed later. Higher action threshold values mean less activities to control and hence a lower effort for the project manager. A specific corrective action is taken when the selected activity shows a delay, and consists of a reduction of the activity delay to half of its original value. Other similar actions have been tested as robustness checks, and have shown no relevant or significant differences.

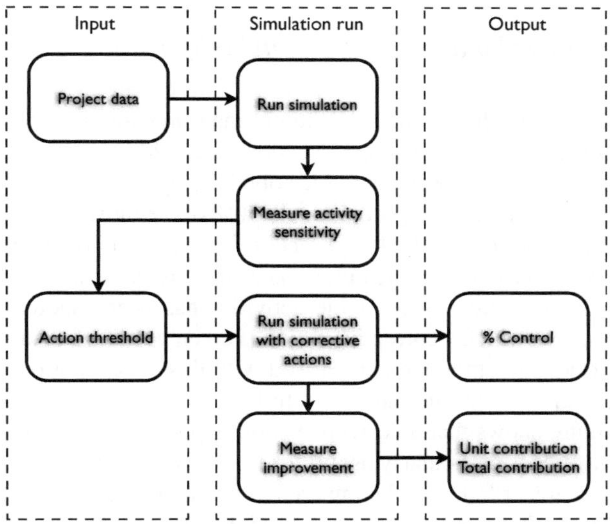

Fig. 5.5 The simulation approach with corrective actions

The three output measures are as follows (each variable has an extra superscript yes/no to refer to the simulation run with or without the corrective action decision making):

- % Control (%C): Percentage of activities in the project network that has been controlled during project tracking (from 0 % (no control) to 100 % (full control)). Consequently, the measure defines the degree of management's attention as the effort the project manager puts in controlling and measuring the performance of the project progress. The % Control and the corresponding corrective actions are triggered by the action threshold, which is set at a value between the minimal (= full control) and maximum (= no control) sensitivity measure value over all activities obtained through the first Monte-Carlo simulation run mentioned earlier.

- Unit contribution (UC): Number of time units (e.g. days) decrease on project duration divided by the total number of time units decrease of all controlled activities as a result of the corrective actions. Hence, this measure calculates the average return of all actions taken on the activities on the total project duration as follows: $\frac{1}{nrs}\sum_{k=1}^{nrs}\left(\frac{RD^{k,\text{no}} - RD^{k,\text{yes}}}{\sum_i (d_i^{k,\text{no}} - d_i^{k,\text{yes}})}\right)*100$.
- Total contribution (TC): % decrease in project duration. This measure calculates the relative contribution of the corrective actions on the total project duration as follows: $\frac{1}{nrs}\sum_{k=1}^{nrs}\left(\frac{RD^{k,\text{no}} - RD^{k,\text{yes}}}{RD^{k,\text{no}}}\right)*100$.

5.3.3 Action threshold = average sensitivity value

Figure 5.6 displays results for the three output measures for activity networks with low, in-between and high SP values. The action threshold is set to the average value of all sensitivity measures for the activities obtained by the first simulation runs. In doing so, the algorithm selects approximately 50% of the most sensitive activities as candidates for the control and tracking process, for which a corrective action will be taken in case of activity delay. The output measures are compared with the random % Control approach where 50% of the activities has been randomly selected as control activities for which a corrective action is taken in case of delay. A horizontal line has been drawn to stimulate comparison with this 50% Control approach. The results of the figure can be summarized as follows:

First, the figure shows that a corrective action approach based on activity based sensitivity measures is particularly useful for low SP value networks. All sensitivity measures show a lower % Control compared to the random approach while both the unit and total contributions are significantly higher. Consequently, the use of these sensitivity measures reduces a project manager's effort while obtaining better results compared to a random control approach.

Second, the figure shows that the CRI and SSI measures seem to keep a relatively low % Control value under various topological structures, while the CI and SI graphs show an increasing % Control as the SP value of the project network goes up. This can be explained by the observation that the CI and SI values are relatively high for more serial project networks (higher SP values) and their standard deviation is relatively low (e.g. for a 100% serial network, all CI values are equal to one, and the standard deviation equals 0). Consequently, it is hard to distinguish between insensitive and sensitive activities and hence more difficult to control the time and effort (% Control) a project manager puts in the tracking process. Note that the effect

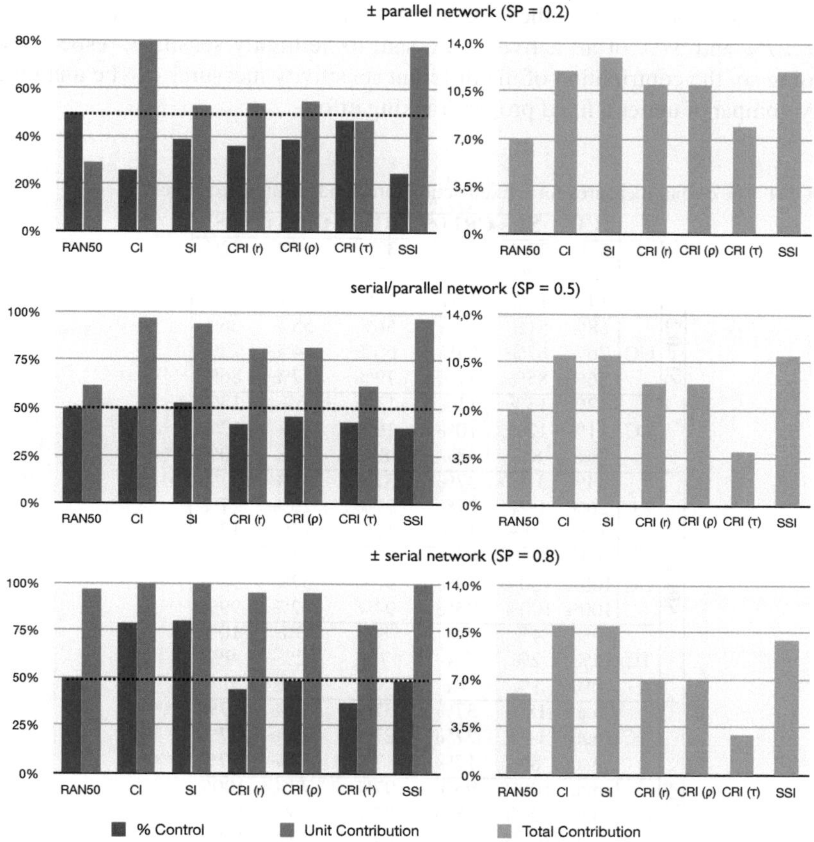

Fig. 5.6 The three output measures for the simulation runs with low, middle and high SP value networks

of the corrective actions, however, is relatively low for the CRI (τ) measure (cf. low total contribution).

5.3.4 Action threshold = x^{th} percentile sensitivity value

In order to investigate the effect of the % Control (i.e. the effort) on the quality of the tracking process for each sensitivity measure, table 5.4 has been constructed for three fixed values of the % Control variable. More precisely, the % Control is kept, to the best possible, constant by using percentiles of the sensitivity measures as action thresholds. More precisely, the 83rd, 73rd and 63rd percentiles are used as action thresholds, such that the % Control variable is set such that approximately 17%, 27% and 37% of all activities are said to be highly sensitive, respectively[3]. In doing so, the contribution of the different sensitivity measures can be unambiguously compared under a fixed project tracking effort.

Table 5.4 The output measures for 3 fixed action thresholds using percentiles

		CI	SI	CRI (r)	CRI (ρ)	CRI (τ)	SSI	
SP=0.2	%C	31%	37%	37%	37%	37%	32%	*
		24%	27%	27%	27%	27%	26%	
		14%	16%	17%	17%	17%	17%	**
	UC	68%	51%	52%	56%	55%	66%	
		76%	67%	64%	66%	64%	72%	
		86%	85%	77%	79%	57%	86%	
	TC	12%	13%	11%	11%	6%	13%	*
		11%	12%	10%	10%	5%	12%	
		6%	8%	9%	8%	2%	10%	**
SP=0.5	%C	34%	31%	37%	37%	37%	37%	
		26%	19%	27%	27%	26%	27%	
		16%	10%	17%	17%	17%	17%	
	UC	98%	98%	84%	87%	63%	97%	
		100%	100%	89%	92%	57%	98%	
		100%	100%	95%	94%	42%	99%	
	TC	3%	4%	8%	8%	3%	10%	
		1%	2%	7%	7%	2%	9%	
		0%	1%	5%	5%	1%	6%	
SP=0.8	%C	n.a.	18%	37%	37%	36%	37%	***
		n.a.	14%	27%	27%	27%	27%	
		n.a.	8%	17%	17%	17%	17%	****
	UC	n.a.	100%	96%	97%	77%	100%	
		n.a.	100%	98%	97%	76%	100%	
		n.a.	100%	98%	98%	70%	100%	
	TC	n.a.	1%	7%	6%	2%	8%	***
		n.a.	1%	5%	5%	1%	6%	
		n.a.	0%	3%	3%	0%	4%	****

The table shows results under three different action threshold values for low, medium and high SP value networks (top, middle and bottom rows, respectively). Each output measure (%C, UC and TC) contains three rows, corresponding to the % Control input values representing action thresholds equal to the 63rd, 73rd and 83rd percentile. Note that the simulated %C values (body of the table, rows %C) are not always exactly equal to their corresponding input values (37%, 27% and 17% for the first, second and third row, respectively) since it is not always possible to find a unique percentile such that exactly 37%, 27% and 17% are above this percentile. This is the case when multiple activities have an equal sensitivity value

[3] As an example, the 83rd percentile of all sensitivity measure values is equal to that value such that approximately 17% of the project activities have a value equal to or higher than this value. For a 30 activity project network, this corresponds to an absolute action threshold of 5 activities.

that is equal to the selected percentile. In this case, all activities with a sensitivity measure value higher than the selected percentile are subject to the activity based tracking and possible corrective actions (i.e. in case of delay) and hence, the real %C value is lower than the corresponding input value. Consequently, the %C variable is controlled in the best possible way, and deviations occur more and more when the sensitivity measures lie close to each other. In the extreme, when the table reports "n.a." for a sensitivity measure, the test was not able to select different values for the percentiles (this is the case when all sensitivity measure values are equal or very close to each other).

Figure 5.7 graphically displays partial results (only the %C and TC output variables) of the table in a graph as follows:

- Low SP and low action threshold: SP = 0.2 and %C = 17% (rows * of the table).
- Low SP and high action threshold: SP = 0.2 and %C = 17% (rows ** of the table).
- High SP and low action threshold: SP = 0.8 and %C = 17% (rows *** of the table).
- High SP and high action threshold: SP = 0.8 and %C = 17% (rows **** of the table).

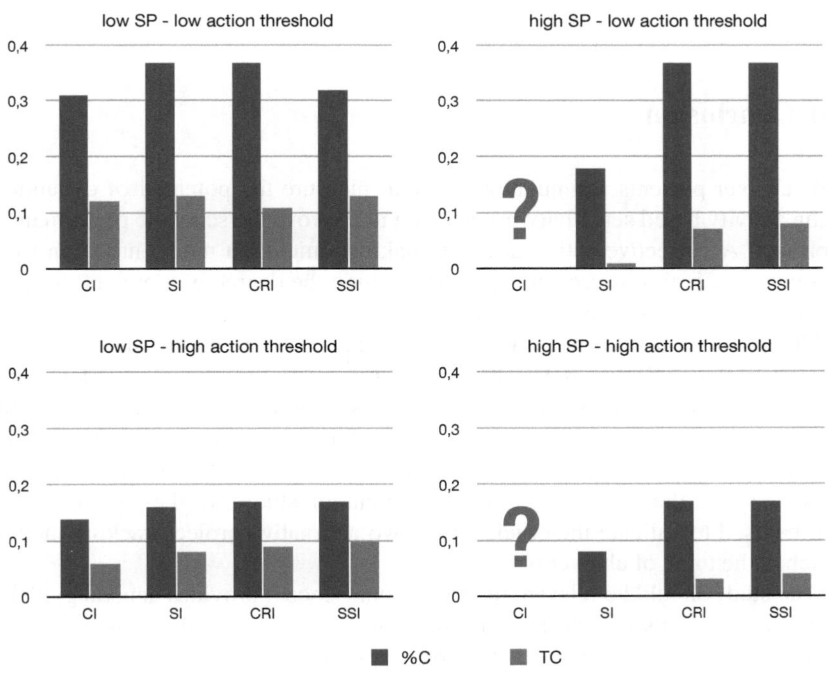

Fig. 5.7 Partial graphical results of table 5.4

The table and the figure are constructed to evaluate the ability of the four sensitivity measures to take timely and effective decisions (i.e. with the lowest effort possible and the highest total contribution) based on accurate information reported by the sensitivity measures. High threshold values (bottom rows for each output measure) are particularly interesting, since they should lead to a small selection (i.e. low % Control) of highly sensitive activities that highly affect the total project duration. The SSI performs best in that respect, followed by the CRI (r) and CRI (ρ) measures, when SP values are low. Even with high action threshold values the total contribution remains relatively high, denoting that a small subset of activities (i.e. leading to a less time consuming tracking approach) is responsible for a high project duration variance. When SP values of projects increase, the CI, SI and CRI (r) measures perform rather poor, as they are not able to select a small subset of activities to take significant corrective actions. Consequently, high action threshold values obviously result in a small selection of activities subject to corrective actions, but it leads to only very small project duration improvements (total contributions drop to 0% or 1% for SP = 0.5 for the CI ans SI). For SP values of 0.8, the CI measure was not able to report different values for the percentiles, as all CI values were high and close to one. The SSI measure, however, shows that even with high action threshold and serial project networks, there is still room to select a small subset of highly sensitive activities, leading to significant contributions when taking the appropriate corrective actions.

5.4 Conclusion

This chapter presents a simulation study to measure the potential of dynamically using activity based sensitivity information to improve the schedule performance of a project. A corrective action decision making simulation model has been run on a large set of fictitious project networks, varying the degree of a project manager's attention on the project during tracking.

The simulation study measures the relevance of four activity based sensitivity measures (the criticality index, the sensitivity index, the cruciality index and the schedule sensitivity index) during project tracking in relation to the network structure, the project manager's effort put in the tracking process and the overall schedule performance improvement when taking corrective actions. The results of the simulation study of this chapter and the two simulation studies of the previous chapter will be used to validate the efficiency of two alternative project tracking methods, which is the topic of chapter 6.

The study should be relevant to practitioners since it provides general guidelines on where the focus of a project manager should be. The results show that project based schedule performance information is particularly useful for serial project networks while more detailed activity sensitivity information is required in a more parallel project network setting. In these cases, management needs a certain feeling of the relative sensitivity of the individual activities on the project objective, in order to restrict the management's focus to only a subpart of the project while still being able to provide an accurate response during project tracking in order to control the overall performance of the project.

Chapter 6
Top-down or Bottom-up Project Tracking

This chapter serves as an integrative summary research study relying on the results obtained from the three previous simulation studies discussed in this book so far. More precisely, a bottom-up and a top-down project tracking approach within a corrective action framework is compared and general summary conclusions are presented. The top-down tracking approach relies on a traditional Earned Value Management (EVM) system investigated in detail in chapter 4, while the bottom-up tracking mechanism makes use of the well-known Schedule Risk Analysis (SRA) method discussed in chapter 5.

6.1 Introduction

Project scheduling, risk analysis and monitoring are crucial steps in the life of a project. The project manager uses the project schedule to help planning, executing and controlling project activities and to track and monitor the progress of the project. A major component of a project schedule is a work breakdown structure (WBS). However, the basic critical path method (CPM) schedules, or its often more sophisticated extensions, are nothing more but just the starting point for schedule management. Information about the sensitivity of the various parts of the schedule, quantified in schedule risk numbers or of a more qualitative nature, offers an extra opportunity to increase the accuracy of the schedules and might serve as an additional tool to improve project monitoring and tracking. Consequently, project scheduling and monitoring tools and techniques should give project managers access to real-time data including activity sensitivity, project completion percentages, actuals and forecasts on time and cost in order to gain a better understanding of the overall project performance and to be able to make faster and more effective corrective decisions. All this requires understandable project performance dashboards that visualize important key project metrics that quickly reveal information on time and cost deviations at the project level or the activity level. During monitoring and tracking, the project manager should use all this information and should set thresh-

M. Vanhoucke, *Measuring Time*, International Series in Operations Research & Management Science 136, DOI: 10.1007/978-1-4419-1014-1_6,

olds on the project level or on lower WBS levels to receive warning signals during project execution. These thresholds serve as triggers to take, when exceeded, corrective actions. These project tracking topics have been discussed in the previous chapters.

This chapter tests two alternative project tracking methods by using two types of dynamic information during project progress to improve corrective action decisions. Information on the sensitivity of individual project activities obtained through schedule risk analysis (SRA) as well as dynamic performance information obtained through earned value management (EVM) will be dynamically used to steer the corrective action decision making process.

The outline of this chapter can be summarized along the following lines. Section 6.2 introduces the two project monitoring methods that will be used throughout this chapter. In section 6.3 the test setting and computational results of the simulation study are discussed in detail. Section 6.4 draws overall conclusions and highlights future research avenues.

6.2 Project scheduling and monitoring

A crucial choice during project tracking is the level of detail and the resulting management effort needed to effectively manage the project tracking process. Hence, the choice of the right WBS level is of crucial importance, and requires a balance between level of detail and ease of project tracking. The importance of the right WBS level has been discussed in literature, and has been mentioned repeatedly throughout the various chapters of this book. Worth repeating is the concluding remark of Lipke et al (2008) who state that "some practitioners of EVM hold a belief that project duration forecasting can be made only through the analysis of the network schedule. These practitioners maintain the understanding and analysis of task precedence and float within the schedule cannot be accounted for by an indicator, and belief that the right WBS level for project tracking is on the activity level." Among others, Book (2006a,b), Jacob (2006) and Jacob and Kane (2004) are authors that express their ideas on EVM project performance measurement and argue that EVM can only be used to measure project time performance as long as the SPI and SPI(t) metrics are used on the activity level, and not on the control account level or even higher WBS levels. However, Lipke et al (2008) also note that detailed schedule analysis is a burdensome activity and if performed often can have disrupting effects on the project team. EVM in general, and the earned schedule method more specific offer calculation methods yielding reliable results on higher WBS levels, which greatly simplify final duration and completion date forecasting. Under this assumption, the EVM metrics are set up as early warning signals to detect in an easy and efficient way (i.e. at the cost account level, or even higher), rather than a simple replacement of the critical path based scheduling tools. This early warning signal, if analyzed properly, defines the need to eventually drill down into lower WBS levels. In conjunction with the project schedule, it allows taking corrective actions on those activities which are in trouble (especially those tasks which are on the critical path).

In this chapter, two extreme tracking methods are taken into consideration, seen from two extreme WBS level starting points. Although they represent a rather black-and-white view on project tracking and monitoring, they can be considered as fundamentally different tracking approaches, both of which can be easily implemented in a less extreme way or can even be combined or mixed during project progress. Figure 6.1 graphically displays the two extreme tracking methods along the WBS level: a project based top-down or an activity based bottom-up project tracking. Details are given along the following lines.

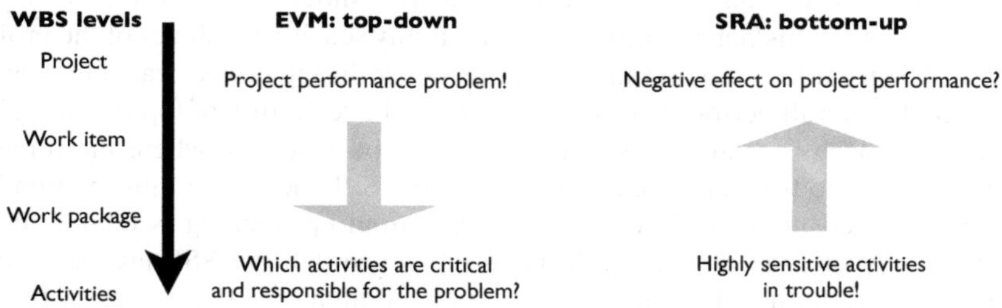

Fig. 6.1 Top-down or bottom-up project tracking approach

Top-down project tracking:

A *project based tracking process* relies on general project performance information that can be used to trigger the activity based corrective action decision making process at the lowest levels of the WBS. EVM provides cost and schedule performance indices (CPI, SPI or SPI(t)) of a project and gives overall project based performance information. A project manager can decide to take corrective actions when a general project performance status drops below a critical threshold (e.g. $SPI(t) < 0.75$). If corrective actions are necessary, the project manager needs to drill down into lower WBS levels (up to the individual activity level) and take the appropriate corrective actions on those activities which are in trouble (especially those tasks which are on the critical path).

Bottom-up project tracking:

An *activity based project tracking* approach (i.e. at lower WBS levels than usually done for the earned value based performance measures) is necessary in case EVM is not applied or does not provide reliable project performance indices, but often demands a stronger control on a larger subset of activities and a continuous critical path based tracking decision process. In this case, individual activity information is required to effectively manage the project tracking and corrective action making decision process in order to improve the overall project performance.

Obviously, the reliability of the EVM based performance measures and forecasts on the project level as well as the SRA based activity sensitivity information are crucial during project tracking and affect the adequacy of the corrective action decision making process. On the one hand, reliable EVM performance measures allow the project manager to restrict the focus on simple project based sanity checks to trigger the often time-consuming critical path based scheduling and tracking process. On the other hand, reliable SRA based activity sensitivity measures provide general guidelines where the focus of a project manager should be. It allows the project manager to restrict his/her attention to only a (highly sensitive) subpart of the project while still being able to provide an accurate response during project tracking in order to control the overall performance of the project. The reliability of top-down project tracking approach is related to the research topic of chapter 4 where the forecast accuracy of methods using schedule performance (SPI and SPI(t)) information has been investigated in detail. The reliability in bottom-up tracking is related to the research study of chapter 5 where it has been shown that the SSI has the highest potential to discriminate between low, medium and high sensitive project activities in a project tracking and monitoring environment.

The next section presents a fourth simulation study that compares the efficiency of both alternative tracking methods. Moreover, the influence of the project network structure, the time uncertainty and the action threshold on the tracking efficiency are tested by means of different simulation runs.

6.3 Simulation 4: A top-down/bottom-up tracking study

6.3.1 Simulation model

This section briefly discusses the settings for simulation test design and the chosen input parameters. Most settings are similar to the settings of the previous chapters, and will not be discussed into detail. Further test-specific details are highlighted throughout the various subsections.

Project data:

The set of projects consists of the 4,100 generated activity-on-the-node networks presented in chapter 4. A crucial parameter used during the project network generation is the Serial/Parallel (SP) indicator, measuring the closeness of a project network to a complete serial (SP = 1) or parallel (SP = 0) network.

Methodology:

The methodology used is Monte-Carlo simulation to generate activity duration and cost uncertainty in a project network, as discussed in the previous chapters. Similar to chapter 5, the simulation model is expanded with a decision making framework with corrective actions to bring late running projects partly back on track. Figure 6.2 gives an overview of the design of the computational experiment, both for the top-down (left) and bottom-up (right) tracking approach. A baseline project schedule is constructed for each project under study using the traditional CPM techniques. Moreover, each project schedule is submitted to a schedule risk analysis, in order to reveal the sensitive parts of the project which will likely have a huge effect on the project objective (time) during project execution. During the Monte-Carlo simulation run, each project is executed as a dynamic process with periodic review periods until the project finish, and its performance is measured on a weekly basis (period t increase) using the earned value calculations. Project tracking and monitoring is done on the project level using schedule performance as a trigger to take corrective actions (top-down approach, left part) or on the activity level using the schedule risk analysis information to select highly sensitive activities as a trigger to take corrective actions (bottom-up approach, right part). If necessary, corrective actions are taken in case the project is in danger, and consists of partly crashing critical or highly sensitive activities to bring the project back on track. At the finish of the project, the tracking efficiency is measured for both tracking methods which consists of a comparison between the effect of the corrective actions on the project duration and the effort (i.e. amount of time and level of detail) the project manager has spent during tracking.

Output measure:

The output indicator is a measure for tracking efficiency, and calculates both the number of times individual activities need to be evaluated for possible activity crashing as well as the number of days the total project duration has been decreased due to corrective actions on individual activities. Consequently, this indicator measures both the effort a project manager puts in the tracking process as well as the effect of his/her actions on the total project duration. The tracking efficiency measure can be given as follows:

$$\frac{RD^{no} - RD^{yes}}{\#\diamond}$$

with

RD^{no}	Real project duration without any corrective action
RD^{yes}	Real project duration with threshold triggered corrective action
$\#\diamond$	Number of times node (a) or node (b) of figure 6.2 is visited

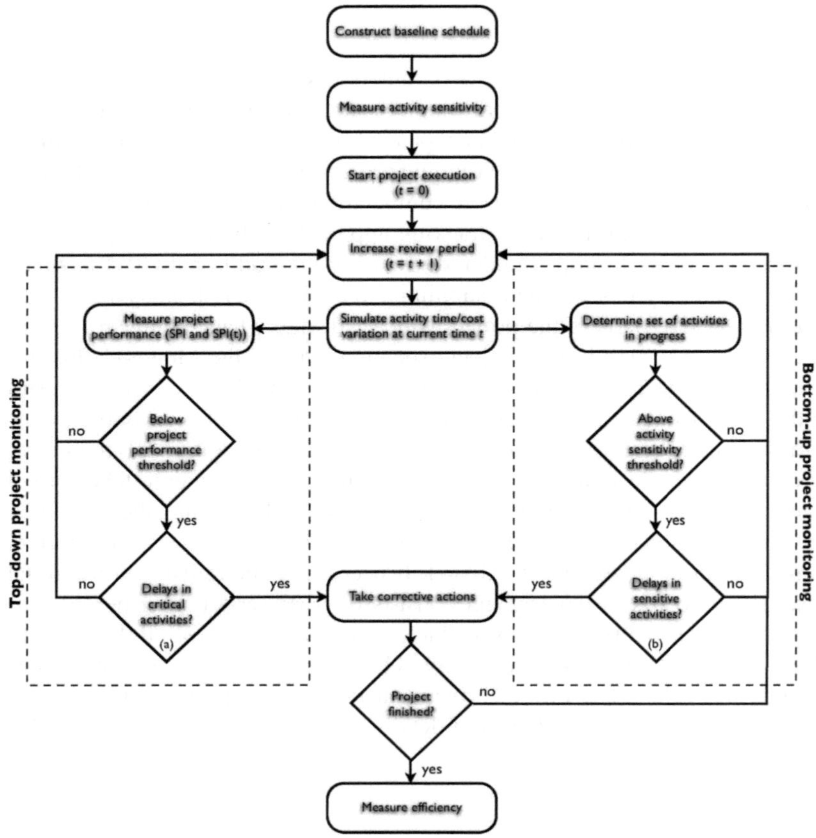

Fig. 6.2 The design of the simulation experiment

The numerator measures the overall effect of all corrective actions on the total project deadline, and compares the real project duration RD^{no} without corrective actions with the situation with corrective actions RD^{yes}. Consequently, the numerator measures the number of time period improvement on the project duration thanks to actions taken by the manager. The denominator measures the project manager's effort as the number of times he/she has evaluated a single activity on its potential delay to take, in case of delays, corrective actions. Hence, this denominator measures the time and effort a project manager spends on checking the lowest WBS level on activity delays. The check is triggered by project performance information (SPI or SPI(t)) for the top-down tracking method, and by activity sensitivity information for the bottom-up tracking method.

Time/cost uncertainty: Random time variation in activity durations is simulated by means of triangular distributions tailed to the right (activity delays) or to the left (activities ahead of schedule), similar to the previous simulations. In the simulation tests, the input parameter mode c of the triangular distribution has always been set to the baseline schedule duration. The lower limit a and the upper limit b vary along the tests, and have been expressed as a percentage of the original baseline duration.

Action threshold: The action threshold approach differs from the two alternative tracking methods. The threshold is set as an absolute value for the EVM top-down tracking method. More precisely, when the performance indicators SPI or SPI(t) drop below a critical threshold, e.g. 0.80, this can be considered as a project delay at the current time instance, and is a trigger to take corrective actions to bring the project back on track. The bottom-up tracking method evaluates the sensitivity of the activities in advance by setting a threshold on the schedule risk analysis values. When an activity's value is larger than the critical threshold (e.g. 90%), this activity needs attention during project tracking. In the current chapter, the threshold has been set as a percentile of all the sensitivity measure values. More precisely, the xth percentile is set as an action threshold, such that the approximately $(100-x)\%$ activities will be subject to intensive control during tracking.

Corrective action: Various corrective action settings have been simulated under various settings. Each time an action is needed, the activity delay will be decreased under a strict limit (e.g. maximum decrease is equal to x% of the activity duration, maximum decrease depends on the cost of the activity, etc.).

In the following subsections, various test results are shown based on different runs under different settings. In these section, only relevant conclusions are illustrated by means of graphs, each time mentioning the specific settings for the random time variation, the action threshold and the corrective actions. Each time, various other scenarios have been simulated as robustness checks, without extra computational results when no relevant changes have been found.

The outline of the following subsections is as follows. In 6.3.2, the effect of the topological structure of a project network on the efficiency of both tracking methods is investigated. Section 6.3.3 measures the influence of time/cost uncertainty on the tracking efficiency. The effect of various fixed and variable action thresholds on the efficiency of both tracking methods is the subject of section 6.3.4.

6.3.2 Effect of the project structure

All project networks have been simulated under time/cost uncertainty scenarios, action thresholds and corrective action settings as described earlier. Figure 6.3 displays results for the bottom-up project tracking method (BU-SRA) and top-down project tracking method using the SPI (TD-SPI) or the SPI(t) (TD-SPI(t)) as an action threshold. The activity uncertainty settings have been chosen to simulate an average activity delay. The action threshold is set as 0.8 for top-down tracking and the 80th percentile for bottom-up tracking, and, when needed, a corrective action is taken as an activity duration decrease of maximum 50% of its baseline duration.

The figure shows the relevance of EVM and SRA information and mainly confirms the results obtained from the previous EVM accuracy and SRA reliability

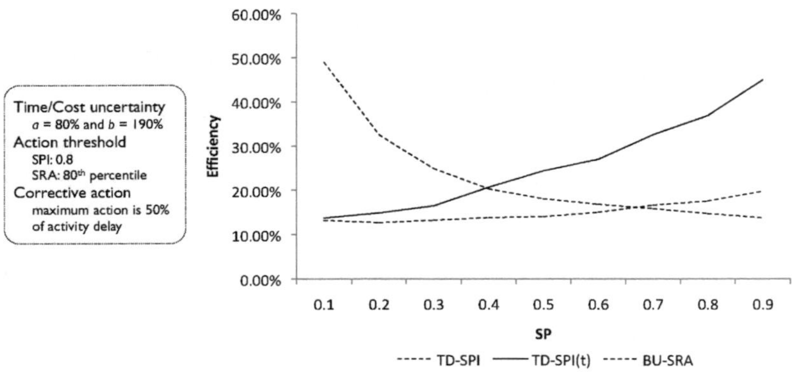

Fig. 6.3 Effect of the project structure (serial/parallel indicator SP) on the bottom-up and top-down project tracking efficiency

studies of chapters 4 and 5, respectively. However, this is the first study where both tracking methods are embedded in a corrective action decision making framework. The results can be summarized along the following lines. First, the graph shows that top-down project tracking using EVM information is particularly efficient for serial networks (high SP values). This observation can be mainly contributed to the reliability of the SPI(t) and (to a lesser extent) the SPI when project activities are executed in series. Second, the figure reveals that bottom-up project tracking is more efficient for parallel networks (lower SP values). These results are in line with the observations and results of chapter 5 where it has been shown that the reliability of schedule risk analysis increases when the presence of parallel activities in the project increases. Finally, the graph also illustrates that the project tracking method using SPI(t) information always outperforms the tracking method based on SPI information, which confirms the results obtained in chapter 4 where it has been shown that the Earned Schedule method outperforms the Planned Value method and the Earned Duration method when forecasting times.

6.3.3 Effect of time uncertainty

In order to measure the effect of activity time uncertainty and the resulting project performance on the efficiency of the top-down and bottom-up tracking methods, all projects have been simulated under various settings, gradually increasing the portion of activity delay. To that purpose, various scenarios have been simulated, where each activity is subject to a triangular distribution where the tail to the right is gradually increased from 150% to 230% in steps of 10%. The runs have been simulated on a subset of the project data set, i.e. project with SP values of 0.2, 0.5 and 0.8.

Figure 6.4 displays the top-down and bottom-up tracking efficiency results for parallel (SP = 0.2), serial/parallel (SP = 0.5) and serial (SP = 0.8) project networks.

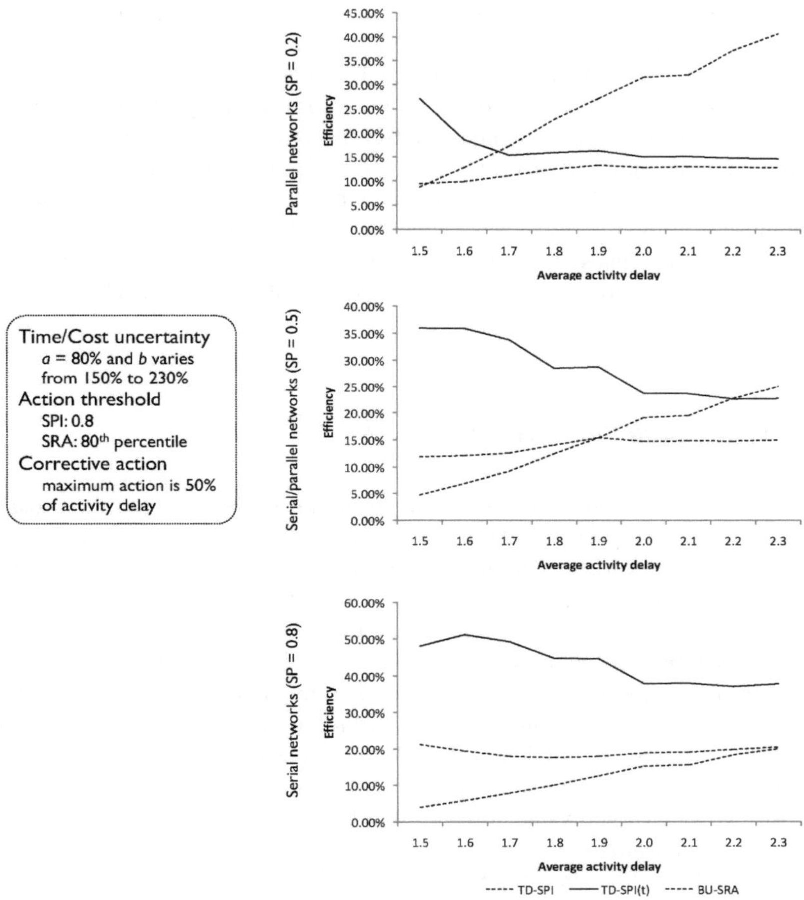

Fig. 6.4 Effect on the average activity delay on the bottom-up and top-down project tracking efficiency

Results for the time uncertainty effect on the tracking efficiency of both methods can be summarized along the following lines. First, the three graphs show that the top-down TD-SPI(t) tracking method is more efficient as the SP value increases and outperforms the TD-SPI tracking method. Both confirm the results from the previous section. Second, all graphs show that the top-down TD-SPI(t) tracking method is more efficient when delays are small. Indeed, in case of small delays, the SPI or SPI(t) measures almost never report an early warning delay signal. In case the metrics do report a delay warning, it's often a correct signal showing a real need for corrective actions. However, in case of more significant delays, the probability that the early warning SPI(t) signal is false increases. Finally, the graphs also reveal that the bottom-up BU-SRA tracking method is not so efficient for small delays. In these cases, most activities are executed according to the scheduled start times, and there is little or no need to take corrective actions to bring the project back on track. However, the SRA bottom-up approach can be considered as a static tracking method, which relies on activity sensitivity information regardless of the project performance. In case of small delays, the BU-SRA method will periodically report tracking warning signals (when the static sensitivity thresholds are exceeded), although there are no delays in these activities and no need for corrective actions. When the size of the activity delays increases, the BU-SRA method will still report identical tracking signals, but will now be able to detect the right activities which will likely have a significant effect on the project duration, which leads to a significant efficiency increase.

6.3.4 Effect of action threshold

In the previous simulation runs, the action thresholds for both the top-down (SPI or SPI(t)) and bottom-up (SSI) methods have been kept fixed throughout the life of the project. However, a variable action threshold which depends on the portion of work already performed might lead to better actions and an increasing tracking efficiency.

In this section, two dynamic action thresholds have been considered above the fixed static action threshold. An increasing action threshold starts with threshold values lower than the fixed threshold value, but gradually increases the threshold value to a predefined maximum value, according to the percentage completion of the project (see figure 6.5). Obviously, a decreasing action threshold value follows an opposite behavior. Note that the percentage completed is measured as the EV / BAC with EV the earned value at the current point in time and BAC the budget at completion. Increasing/decreasing action thresholds have an opposite meaning for bottom-up and top-down tracking methods, as follows:

- Top-down: An increasing SPI(t) action threshold value denotes a low attention at the start of the project, which gradually increases to a sharpened attention as the project work completes. This is a situation where SPI(t) delay warnings can often be ignored at the beginning of the project, since they are often based on insufficient or inadequate data, but since the quality improves towards the end of the project, the performance metrics get a higher attention. A decreasing SPI(t) action threshold value seriously considers the warning signals from the very beginning, since delays might have an effect on the rest of the project, and assumes a weakened project performance attention as the project work goes on.
- Bottom-up: The lower the SSI action threshold, the more activities are considered as sufficiently sensitive to be subject to the tracking approach. Consequently, an inreasing SSI action threshold value denotes a sharp attention from the start

and assumes that the attention can be decreased as the project work goes on. A decreasing SSI action threshold assumes the opposite behavior.

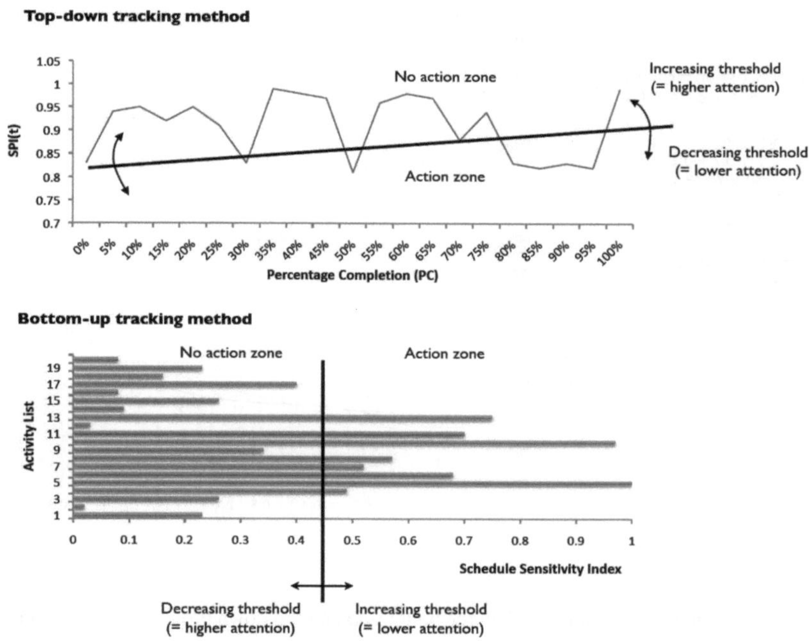

Fig. 6.5 Static (fixed) and dynamic (increasing and decreasing) action threshold values

Figure 6.6 displays the results for the BU-SRA and TD-SPI(t) tracking methods under three different action thresholds. Results for dynamic versus static action threshold simulation runs can be summarized along the following lines. First, the figure confirms the previous tests of section 6.3.2 that the SP indicator is responsible for the selection of the best performing tracking method. Second, the top figure shows that bottom-up tracking using SRA is particularly useful in combination with decreasing action threshold values. In this case, the tracking method assumes a rather weak tracking effort at the start of the project, while this is gradually increased towards the end of the project, considering more and more activities as highly sensitive and subject to further control. Finally, the top-down tracking approach shows an opposite behavior. The graph shows that decreasing action threshold values lead to the best results, which means that the early project stages are crucial and deserve more attention than the late project stages. Early warnings signals need to be interpreted quickly to take the correct actions that might affect future project performance.

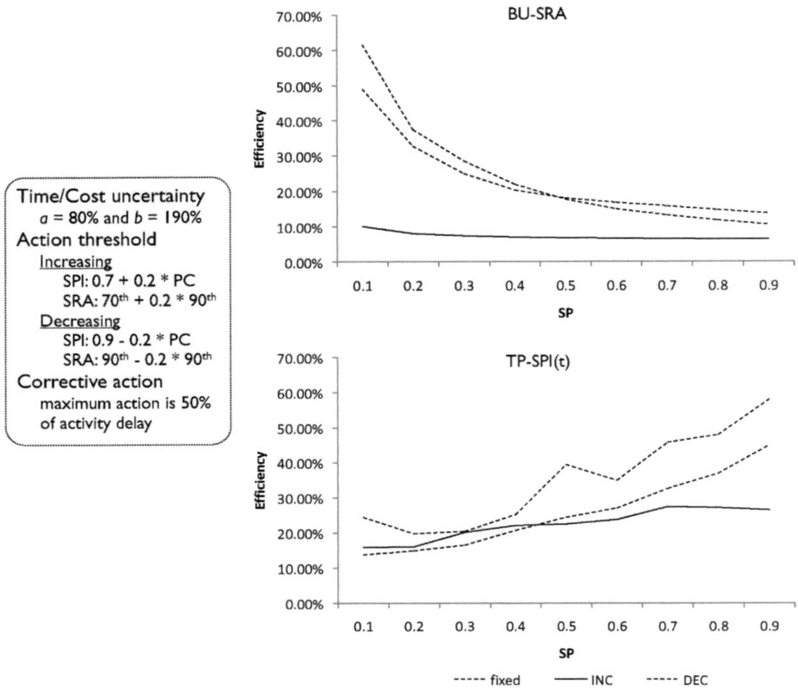

Fig. 6.6 Effect on the action threshold on the bottom-up and top-down project tracking efficiency

6.4 Conclusion

In this chapter, two alternative project tracking methods are compared and validated on the set of fictitious project data using Monte-Carlo simulations. The top-down project tracking method relies on EVM project performance data that are used as early warning signals and triggers to the need for corrective actions, as discussed in chapter 4. The bottom-up tracking method is based on schedule risk analysis that reveals sensitivity information of each activity and hence the need to focus on only the highly sensitive parts of the project, as discussed in chapter 5.

Each project network is simulated under various settings, each time measuring the efficiency of both tracking methods as a combination of the project manager's effort during project tracking and the results of actions on the project duration. A computational experiment has been set up under various settings, each time measuring the tracking efficiency of both alternative project tracking methods. The tests contain three subtests. A first run tests the effect of the project network structure on the tracking efficiency, a second run evaluates the effect of the project performance, measured as the average percentage activity delay, on the tracking efficiency while a last run investigates the effect of static as well as dynamic action thresholds as a function of the project percentage completion on the tracking efficiency.

The simulation study confirms the results found in the previous chapters. The tracking efficiency is high for a top-down project tracking approach using EVM information when the project contains many serial activities. This is completely in line with the forecast accuracy study of chapter 4. The tracking efficiency is high for a bottom-up tracking approach using SRA information when the project contains many parallel activities. These results confirm the results and conjectures made in chapter 5. Although this fourth simulation study sheds a rather black-and-white view on project tracking, it provides results that should be relevant to practitioners

Chapter 7
ProTrack: A Software Tutorial

ProTrack (acronym for Project Tracking) is a project scheduling and tracking software tool developed by OR-AS to offer a straightforward yet effective alternative to the numerous project scheduling and tracking software tools. The software has been built based on the results of the research studies discussed in this book and the many discussions with practitioners using earned value management. The scheduling and tracking approach is based on the current best practices of earned value management and the novel concepts introduced and tested in this book.

This chapter highlights the main characteristics of ProTrack that are novel compared to traditional project management software and presents an overview of the features discussed throughout the various chapters in this book. Section 7.1 reviews the main project scheduling options available in ProTrack. In section 7.2, the main earned value/schedule based project tracking possibilities are highlighted and discussed. Section 7.3 presents three novel optional engines that can be added to ProTrack to extend the capabilities of the software to automatic project generation, various simulation runs and time forecast accuracy calculations. In section 7.4, a demo experiment illustrates the use of ProTrack on a set of 9 generated projects.

7.1 Project scheduling with ProTrack

The target of project scheduling is to construct a timetable where each individual activity receives a start time and a corresponding finish time within the predefined precedence relations and the various predefined activity constraints. The scheduling process is based on the traditional critical path based forward (to create an earliest start schedule) and/or backward (to create a latest start schedule) project scheduling calculations aiming to construct a project schedule with a minimal project lead time. In the remainder of this section, details on the exact use of the precedence relations, the activity constraints and the forward/backward scheduling options are described in order to create a project baseline schedule.

M. Vanhoucke, *Measuring Time*, International Series in Operations Research & Management Science 136, DOI: 10.1007/978-1-4419-1014-1_7,

7.1.1 Precedence relations

A precedence relation between two activities can be one out of four types, in line with the traditional software and current state-of-the-art literature. More precisely, a choice has to be made between a Start-Start (SS), a Start-Finish (SF), the default Finish-Start (FS) and a Finish-Finish (FF) relation with a minimal[1] time lag equal to (default) or different from zero. Figure 7.1 shows the four types of precedence relations between a pair of activities with each a positive time lag displayed by the dashed lines between the activities.

Fig. 7.1 The 4 types of precedence relations

7.1.2 Activity constraints

Activity constraints can be imposed when there is a need to manually control the start or finish of an activity. ProTrack contains three types of activity constraints, each having a start and a finish version, as follows:

- *Ready dates* imply earliest start or finish times on activities and hence force the activity to start/finish no earlier than the defined time instance. These constraints are known as ready start time (RST) or ready finish time (RFT).
- *Due dates* imply latest start/finish times on activities and force activities to start/finish no later than a predefined time instance. In ProTrack, these constraints are referred to as due start time (DST) or due finish time (DFT).

[1] In literature, a difference has been made between *minimal* and *maximal* time lags. When only minimal time lags are present in project networks, it is referred to as the precedence diagramming method (Moder et al, 1983), whereas project networks with both minimal and maximal time lags are referred to as generalized precedence relations (Elmaghraby and Kamburowsky, 1992). In ProTrack, only minimal time lags can be incorporated in the project network.

- *Locked dates* imply a fixed time instance and force the activity to start/finish on a predefined time instance, known as locked start time (LST) or locked finish time (LFT).

Obviously, the use of activity constraints increases your own control to the project schedule but leads to a flexibility decrease for the project scheduling algorithm. Although it can be generally recommended to restrict the use of activity constraints to prevent the construction of a rigid project schedule, ProTrack provides a project based choice of hardness applicable to all activity constraints, varying between a "hard" mode, a "moderate" mode, a "soft" mode or a "forward" mode. These hardness options influence the result of user interventions (e.g. a manual activity shift in time, adding a constraint or precedence relation, changing an activity duration, etc.) or software intervention (e.g. rescheduling the baseline schedule, update of tracking information, etc.) on the project schedule.

- *Hard constraints* need to be followed at all times, and hence, when an intervention leads to a possible violation of activity constraints, ProTrack will return to the previous schedule and undo the infeasible user or software intervention.
- *Moderate constraints* cannot be violated due to interventions, and when infeasibilities tend to occur, the precedence relations will be overruled by allowing a certain degree of overlap between project activities. Consequently, a moderate activity constraint has a higher priority than a precedence relation between activities.
- *Soft constraints* can be violated by user or software interventions at any time, but the software will try to prevent the total number of violations. To that purpose, each activity constraint needs a certain weight as a measure of importance to satisfy the constraint. In case of violations, a report with the violated constraints will indicate where the overridden requirements occur.
- In the *forward mode*, only forward activity constraints are taken into account. Consequently, all activity ready times are explicitly taken into account, while locked times and due dates are often ignored. More precisely, locked times are treated as ready times, which means that they will only be satisfied unless it is not possible due to predecessor activities. Due dates are completely ignored and will possibly be violated by user or software interventions at any time. However, a report with the violated constraints will indicate where the overridden requirements occur. This mode is particularly useful during project simulation and project tracking, where activity constraints set during the construction of the baseline schedule can often not be satisfied due to delays in predecessor activities. Moreover, too much constraints often bias the simulation results due to the influence and rigidness of the activity constraints.

While the moderate mode (always respect the activity constraints at the cost of violation of precedence relations) and the forward mode (only obey the ready times and partially obey the locked times) are intuitively clear and straightforward, the hard (always respect the activity constraints, unless it is impossible) and soft mode (try to respect the activity constraints in the best possible way) are completely new in project scheduling software, and deserve further explanation.

In the hard mode activity constraint option, interventions often lead to infeasible project schedules due to one or more activity constraints that cannot be satisfied. As an example, think of an activity shifted further in time, such that a due date constraint of a successor activity will be violated. Obviously, the software will report an error message indicating the infeasibility problem of the intervention, and will return to the previous schedule saved before the intervention. However, ProTrack offers an extra ability to allow the user to satisfy the set of activity constraints, if possible, which can be useful when the user makes a switch from a certain constraint mode (forward, soft or moderate) to the hard mode. In these cases, it is possible that the constraint mode switch leads to infeasible schedules given the start times of the start schedule, but might lead to a feasible earliest start schedule (i.e. by undoing all previous activity shifts). In the latter case, ProTrack will report an infeasible activity constraint switch and will allow the user to either undo the switch or to continue with an earliest start schedule satisfying all the hard activity constraints.

In the soft mode option, a project schedule is constructed from scratch in order to satisfy the activity constraints in the best possible way. ProTrack makes use of the efficient recursive search procedure developed by Vanhoucke et al (2001) to solve the so-called weighted earliness-tardiness project scheduling problem. This procedure always reports a project schedule such that the weighted deviation between each activity start/finish time and its corresponding constraint is minimized. The schedule is each time constructed from scratch, which means that simple forward or backward activity shifts would normally have no effect on the resulting schedule. Indeed, shifting an activity would lead to an automatic undo operation since the start schedule, before the shift, was already the best possible soft mode schedule. In order to avoid this, ProTrack will set the new start or finish time on the shifted activity as a new constraint. Consequently, the construction of a baseline schedule in this soft mode option is referred to as *constraint based scheduling* since activity changes (shifts, duration increases, etc.) imply an automatic addition of activity constraints, as follows:

- Forward shift (= towards the project finish): set a ready time at the start of the new activity starting time (RST)
- Backward shift (= towards the project start): set a due date at the finish of the new activity finish time (DFT)

Figure 7.2 shows the four activity constraints hardness options on a fictitious three activity project with two precedence relations between activities 1 and 3 and between activities 2 and 3. The first column shows the result of the introduction of three activity constraints into the earliest start schedule shown at the top left corner. Activity 1 gets a ready time at its current start time, activity 2 gets a ready time somewhat later than its current start time and activity 3 gets a due date which is set later than its finish time. Since the introduction of these activity constraints does not lead to constraint conflicts, the "hard" mode, "moderate" mode, "soft" mode and "forward" mode all show an identical Gantt chart where the second activity start time has been shifted automatically toward the project finish. The right column shows the result of a manual user intervention where activity 1 has been shifted forward

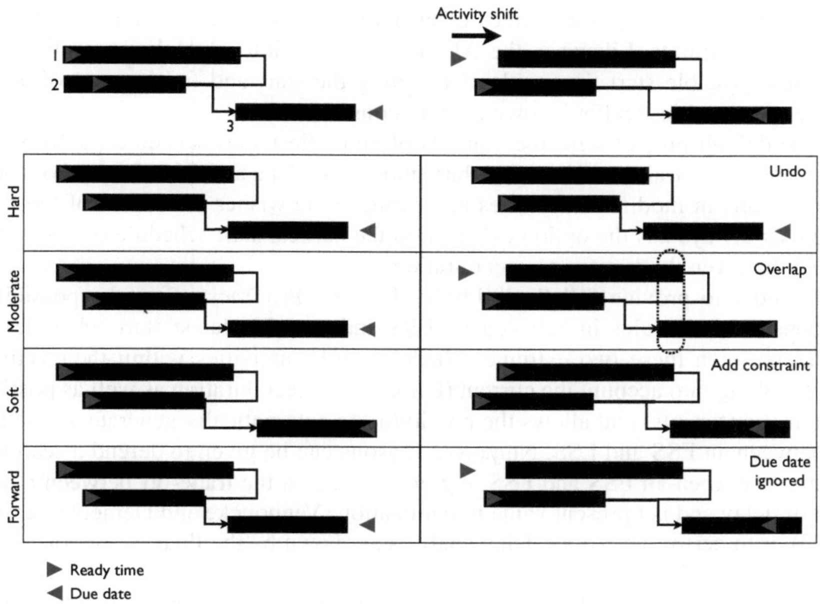

Fig. 7.2 Activity constraint hardness options in ProTrack

in time. The "hard mode" option detects a conflict between the constraints, which leads to an undo operation to cancel the shift. The "moderate mode" option always obeys the activity constraints, and hence, schedules an overlap between activities 1 and 3 to satisfy both activity constraints. The "soft mode" option always respects the precedence relation, and tries to minimize the deviation between the predefined activity constraint and the start/finish time of the corresponding activity. Moreover, a shift of activity 1 results in the introduction of a new ready time constraint. The figure shows a possible schedule, where the constraint of activity 3 is violated in favor of the first activity constraint (the chosen solution depends on the weight of each activity constraint). Alternatively, the scheduler could violate the first activity constraint, in favor of the third activity constraint. The "forward mode" option only obeys the forward activity constraints (ready times and locked times treated as ready times) and violates the due date constraint of the third activity due to the shift.

7.1.3 Earliest/Latest start schedule

While an activity constraint is an optional field for each activity, the as-soon-as-possible (ASAP, default option) or as-late-as-possible (ALAP) option is required. An ASAP activity schedules the activity on its earliest possible start time without affecting the start and finish times of all its predecessor activities and/or its own activity constraints. Likewise, the ALAP activity option schedules the activity on its latest possible start time without affecting the start and finish times of all its successor activities and/or its own activity constraints.

The default project schedule consists of an earliest start schedule (ESS) where the activity constraints are satisfied (hard mode). Shifting activities forward or backward results in modified schedules according to the wishes and needs of the user. The user always has the option to return to the earliest start schedule or latest start schedule, given the current project duration.

In order to provide full flexibility to the user, ProTrack offers the possibility to construct schedules in between an ESS and an LSS (latest start schedule). A slider between these two extreme schedules shifts activities within their activity slack (taking into account the current (buffered) project duration as well as possible activity constraints) and allows the possibility to automatically generate a schedule in between an ESS and LSS. Numerous reasons can be given to defend a search to balance between an ESS and LSS, e.g. to balance on the trade-off between a high risk of delay and net present value maximization (Vanhoucke and Demeulemeester, 2003) or to perform a risk of delay analysis as shown by the float factor concept of Tavares et al (1999).

Figure 7.3 illustrates the concept on a three activity project with an extra option to extend the project schedule with a user-defined project buffer. Obviously, the role of the project buffer is to extend the length of the total slack or float of each activity and serves as a safety time above the critical path length and a deadline promised to the customer of the project. By changing the schedule from an ESS (float factor = 0%) to an intermediate (float factor = 50%) and an LSS (float factor = 100%), each activity is shifted further in time within its calculated slack, without violating the total project deadline defined by its critical path based duration extended with a possible project buffer (PB).

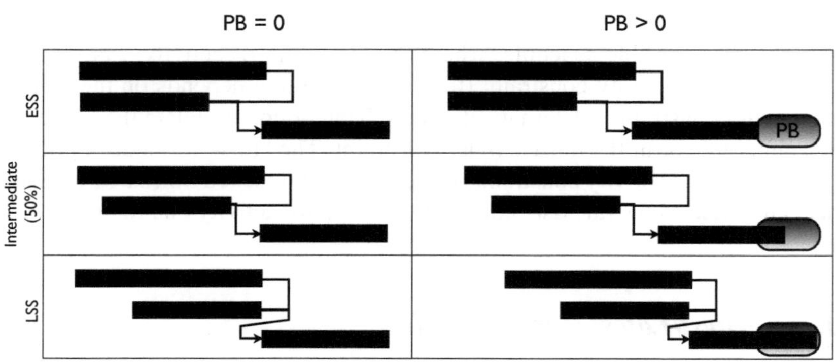

Fig. 7.3 Choices between an ESS and an LSS (from 0% to 100%)

7.1.4 Baseline schedule

The development of a detailed baseline schedule for measuring schedule performance (time, cost, resources, etc.) as well as to perform risk analyses to detect a project's most sensitive parts is a crucial step in project scheduling and tracking. The main purpose of the baseline schedule is to act as a point of reference for project tracking (see chapter 4) and schedule risk analysis (see chapter 5). In ProTrack, the baseline schedule can be saved by the user, or will be automatically saved from the moment the first tracking period schedule is constructed (see section 7.2). ProTrack can save as many baseline schedules as the computer memory allows but makes a strict distinction between the construction of a baseline schedule and the tracking process during project execution. Figure 7.4 illustrates this distinction and shows that making changes to the original baseline schedule during project tracking is only possible by re-baselining the original baseline schedule (i.e. overwrite the current active baseline schedule) and adding it to the current and possible future tracking periods (see section 7.2). In doing so, each tracking period schedule is linked to a single and unique baseline schedule, resulting in a unique and unambiguous point of reference for earned value metric calculations. However, it should be noted that one should be careful with unwise re-baselining decisions, since this involves a change in the tracking point of reference, and hence, previous tracking results are no longer comparable with the current tracking progress state.

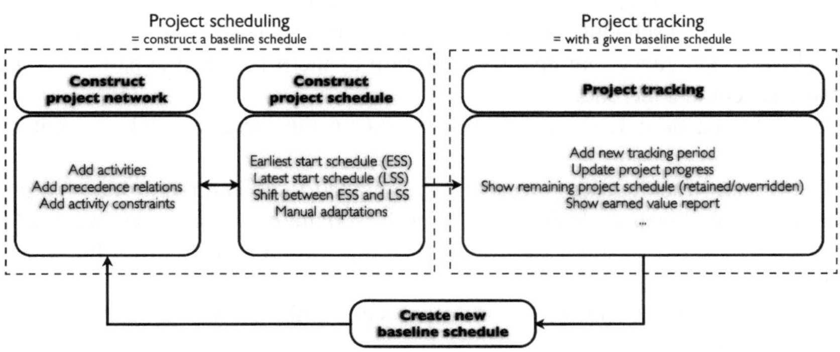

Fig. 7.4 Project scheduling and tracking in ProTrack

7.2 Tracking progress with ProTrack

Project monitoring or tracking is the process performed to observe project execution in order to identify potential problems and/or opportunities in a timely manner such that corrective actions can be taken when necessary. The key benefit is that the current project status is observed on a regular basis, which enables the calculation of the project performance variance that is equal to the gap between actual performance and the baseline schedule.

Since the current project performance is measured by variances from the project management plan, the baseline schedule plays a central and unambiguous role during the project tracking process. It has been mentioned earlier that ProTrack employs a strict definition of a unique baseline schedule for each tracking period, such that it is always clear during the complete project execution what the active baseline schedule is. ProTrack allows the user to define multiple tracking periods in order to get a clear view of the progress of the project over time. Each tracking period is linked with a unique status date and a corresponding baseline schedule, and measures the current progress up to the status date of the project compared to the active baseline schedule. When multiple baseline schedules have been constructed, the user needs to select a single baseline schedule for each tracking period. Consequently, this strict baseline schedule definition guarantees a clear and unambiguous interpretation of the current performance (i.e. the current performance compared to the *active* baseline schedule at the status date) and guarantees that only one baseline schedule is active at the same time.

The ProTrack tracking options mainly follow the principles and techniques discussed throughout the various chapters of this book, and are briefly summarized as follows:

- Earned value: automatic calculation of the standard EVM key metrics and extensions to earned schedule project tracking and performance measurement.
- Schedule adherence: automatic calculation of the p-factor to dynamically measure schedule adherence.
- Time forecasting: EVM forecasting based on the three methods presented in chapter 4.
- Gantt chart tracking: a choice between retained logic and overridden logic (all intermediate levels inclusive) to predict the remaining work shown in a tracking Gantt chart. This Gantt chart can replace the current baseline schedule (i.e. rebaselining) when necessary.
- Reporting: a wide range of flexible reports customized with various logos can be easily made.

In section 7.2.1, the different input parameters that are necessary to automatically calculate all earned value and earned schedule metrics are briefly discussed. Section 7.2.2 explains the retained/overridden logic options in order to allow Gantt chart project tracking. Section 7.2.3 briefly illustrates the flexibility in reporting the performance status of a project at a current point in time.

7.2.1 Earned value/earned schedule

ProTrack's project tracking can be done by creating a new tracking tab for each review period. Each tab is linked to a unique status date and baseline schedule, and multiple tabs allow the user to get back in time and review previous tracking periods.

The earned value module in ProTrack contains the classic earned value key parameter calculations as well as the novel earned schedule metrics and allows the use of all performance measures (CPI, SPI and SPI(t)) and the EAC(t) forecasting indicators as presented in chapter 1. Moreover, the p-factor has also been included as a simple way to dynamically measure schedule adherence during the progress of the project.

Correct earned value based calculations depend on the quality and accuracy of the input parameters, which can be summarized along the following lines:

- Actual Start: actual starting time of an activity on or before the status date which might differ from the original baseline start time.
- Actual Duration (AD): the time spent on an activity between its actual start time and the current status date. Hence, the actual duration can be smaller than, equal to or larger than the time span between the actual start and the current status date, and is used to represent the workload already spent on this activity until now.
- Remaining Duration (RD): a forecast of how many time units (hours, days, ...) an activity will need from the status date forward to finish its remaining work.
- Actual Cost (AC): the actual cost for the work already done. Normally, this actual cost should have a clear relation with the workload already done by the various resources as given by the actual duration, but the user is free to adapt the actual cost to any monetary value.
- Remaining Cost (RC): a forecast of how many additional monetary resources (euros, dollars, ...) an activity will need to finish the portion of remaining work.
- Percentage Completed (PC): the portion of the total workload that is estimated to be finished. The earned value metric EV is based on this estimate as EV = PC * BAC. Note that neither the actual/remaining duration nor the actual/remaining cost needs to have a link with the PC estimate and will not be explicitly used for the calculation of the EV metric.

Figure 7.5 shows an activity with a planned duration of 10 weeks and a budget at completion BAC = € 100. The second part of the figure ("project tracking") shows a project tracking Gantt chart with two user defined input parameters, actual start and status date, and the default values reported by the software for all other tracking parameters. The default formulas are based on the manual input of the status date and actual start. Obviously, the software user needs to be careful with default values! Beware that these values do often not reflect reality, and require manual adaptations.

1. Actual Duration = status date - actual start
2. Remaining Duration = Planned Duration - Actual Duration
3. Actual Cost = BAC / PD * Actual Duration
4. Remaining Cost = BAC / PD * Remaining Duration

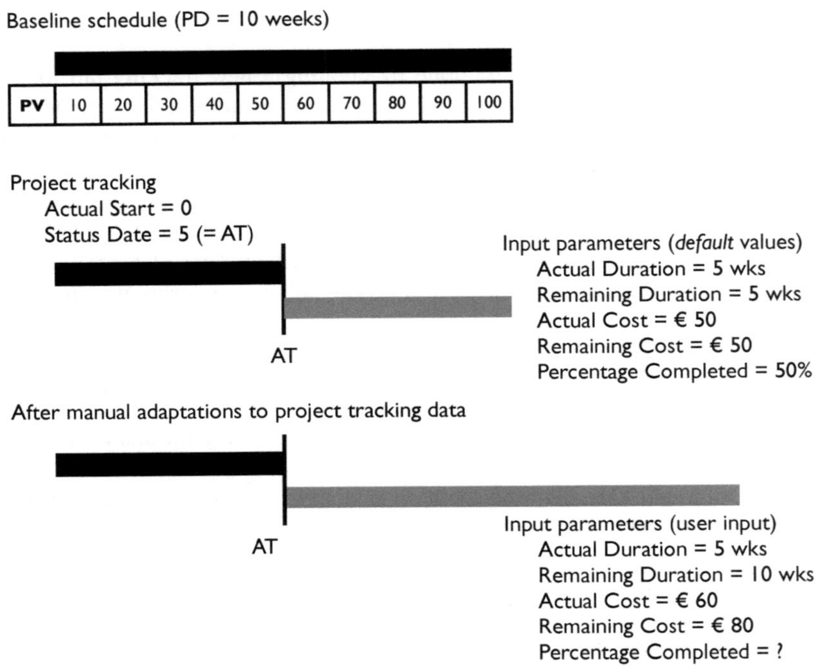

Fig. 7.5 Project tracking input parameters

5. Percentage Completed = Actual Duration / (Actual Duration + Remaining Duration)

The bottom part of the figure ("manual adaptations") shows an update of the project tracking Gantt chart by modifying the tracking input parameters manually by the user. The crucial question arises how to measure earned value, defined as PC * BAC. The percentage completed is used to calculate the current reporting period earned value for each project activity. The challenge for this approach is to define a value for the percentage completed, since estimating percent completed is a subjective exercise open to abuse by the optimistic and unscrupulous. There is no doubt that having a value that represents truth and reality is an important and crucial step in EVM tracking and reporting. Therefore, companies often evolve to a practice of having employees estimating their percent complete at each status period and having team leads and/or project managers validating and possibly correcting the entries. Having some amount of oversight involved often helps to make the estimate closer to reality. Nevertheless, whatever method is used, estimates are and will always remain subjective and open for abuse, as illustrated by the numerous discussions on the internet on "how to lie with earned value?". Table 7.1 gives an overview of different EVM measurement methods as a summary given and discussed by Fleming and Koppelman (2005). ProTrack only incorporated the "percentage completed" estimate.

Obviously, a software tool will never be able to replace the human expertise necessary to define accurate and realistic estimates on the percentage completed. However, fast and easy heuristics might help and are incorporated in ProTrack.

1. % Completed = manual input, i.e. an estimate made by the ProTrack user (best method!)
2. % Completed = AC / (AC + RC) = 60 / 140 = 42%
3. % Completed = AD / (AD + RD) = 5 / 15 = 33%

7.2.2 Retained or overridden logic

One of the primary tasks during project tracking is that the baseline schedule is periodically updated to reflect the actual progress of the work done and to present a realistic forecast of the remaining work. Since work is often performed out-of-sequence, the original logic captured by the precedence relations between activities is often violated. This situation can cause unrealistic deviations between the baseline scheduling logic and the project tracking Gantt chart, and often leads to unnecessary adaptations and modifications to the baseline schedule. ProTrack has two options to handle out-of-sequence progress during the tracking phase, as follows:

- Retained logic assumes that the original precedence relations are still valid, even when activity overlapping during progress has taken place. This logic respects all precedence relations of the remaining work, but often leads to unrealistic long project duration forecasts.
- Overridden logic assumes that an activity that started with a certain overlap will violate the original precedence relation logic completely. This logic assumes that the remaining work of an activity in progress can be done without being affected by its incomplete predecessor activities, but it often leads to unrealistic short project duration forecasts. This logic is also known as out-of-sequence progress.

Figure 7.6 illustrates the two extreme out-of-sequence logics on a simple three activity project, as well as a baseline schedule and an in between option that will be discussed hereunder.

While the option between the retained and overridden logic is standard in most commercial software tools, ProTrack has extended this option with two extra features:

Overridden/Retained logic percentage: ProTrack has the option to construct a remaining schedule in between the overridden and retained logic modes, as shown by the "in between" schedule of figure 7.6. Motivated by the observation that the Gantt chart of the remaining work is nothing more but a schedule forecast, the option to shift between the two extreme logics allows the user to fine-tune the forecast according to his/her own wishes. A simple slider, similar to the ESS/LSS slider

Table 7.1 Earned value measurement methods (Source: Fleming and Koppelman (2005))

Measurement Method	Description	Notes
Weighted Milestone	Useful for short-span tasks. Weighted budget amounts are applied to milestones distributed across the duration of the activities. As the milestone is reached, the budget amount is earned.	
Fixed Formula (e.g. 0/100; 25/75; 50/50)	Useful for detailed short-span tasks. An amount ranging from 0 to 100 percent of the task's budget is earned when the task begins. The remaining percentage is earned when the task is complete.	These measures are typically used to track earned value on non-recurring tasks.
Percentage Complete Estimates	A "subjective" estimate of the percentage of work completed is used for the earned value. Requires well-defined work packages and guidelines to determine an accurate percent complete value.	Supported by ProTrack.
Percentage Complete and Milestone Gates	Uses a combination of weighted milestones and percent complete. Subjective percent complete estimates are allowed up to a specific ceiling associated with each milestone. Advancement past the milestone is not allowed until tangible criteria have been met, hence the term "gate".	
Equivalent Completed Units	Useful for extended duration and repetitive tasks. The overall project is divided into distinct units of accomplishment. Earned value is computed by summing the units complete.	
Earned Standards	Standards of performance based on historical cost data, time and motion studies, are used to compute the earned value on a given task. Management consensus determines which standard is ultimately used.	These measures are typically used to track earned value on either non-recurring or recurring tasks.
Apportioned Relationship to Discrete Work	The earned value for apportioned tasks is a summary of the earned value measurements made on the work to which it is related. Schedule variances in the apportioned work are usually identical to schedule variances in its related work. However, cost variances in the apportioned work are substantially different from the related work because of the dynamics of actual costs.	This measurement can employ any of the above six methods.
Level of Effort	Generally, level of effort tasks are those that support the overall project. Because these tasks are more time driven than task driven, whatever is set as the planned value is always the earned value, regardless of what work was done.	This method is generally not recommended to track earned value.

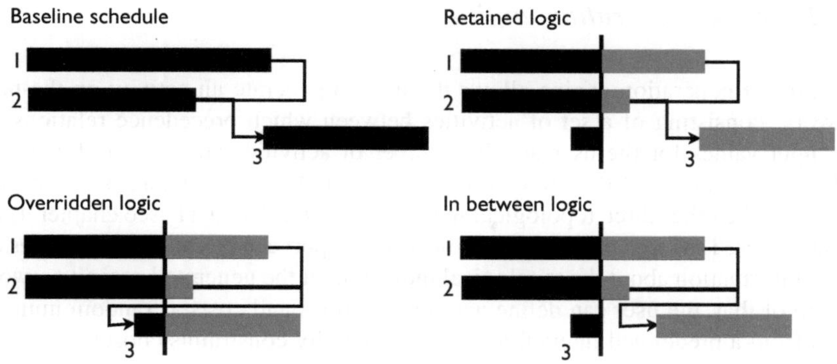

Fig. 7.6 Retained and overridden logic (and in between) options in ProTrack

mentioned earlier, gradually shifts the tracking Gantt chart from an overridden logic to a retained logic situation.

Buffer management: This overridden/retained logic slider plays an important role in the buffer management approach set during the scheduling phase (see section 7.1.3). During the project scheduling phase, a buffer is set as a well-considered choice of a certain degree of safety time on top of the minimal critical path based project duration. This buffer acts as a simple project tracking dashboard, since project tracking leads to possible buffer consumption and hence visualizes the likelihood of violating the predefined deadline. The options between the overridden and retained logic obviously affects the buffer consumption and can be used to predict and influence the estimated project finish.

7.2.3 Project reports

Reporting lies in the heart of project tracking and performance measurement. The use of flexible reports to illustrate EVM output measures is crucial for effective project tracking. Tables as well as graphs, with exporting options to Microsoft Excel are standard functions implemented in ProTrack.

7.3 ProTrack engines

ProTrack can be extended to three optional engines that have been discussed throughout the previous chapters, i.e. the project generation engine, the simulation engine and the time forecasting engine.

7.3.1 Project generation engine

The project generation engine allows the user to generate an activity-on-the-node network, consisting of a set of activities between which precedence relations exist. Input values for the user are the number of activities and a value for the serial/parallel indicator SP $\in [0,1]$ to define the topological structure of the project network. The other three topological indicators (AD, LA and TF, see chapter 4) are not considered as input parameters but instead, a report can be called which gives detailed information about the topological indicators of the generated project network. On top of that, the user can define activity durations and costs as random numbers taken from a predefined interval and a set of activity constraints chosen at random from the three constraint types defined earlier. The generation process is based on the generation principle used for the RanGen1 (Demeulemeester et al, 2003) and RanGen2 (Vanhoucke et al, 2008) generators, which have been shown to be full alternatives to existing project network generators in literature (for an overview, see section 4.1.1).

Figure 7.7 shows the project generation engine approach in ProTrack with the input network and activity parameters as described in this book. The number of activities as well as the SP indicator of chapter 4 are necessary input to determine the topological structure of the network. Activity information contains time and cost information for individual activities, as well as the option to randomly assign activity constraints across various activities in the project network. ProTrack will automatically generate a project network and a corresponding baseline project schedule, and allows the user to call for more project information (the values for the three other topological indicators AD, LA and TF, inclusive).

7.3.2 Simulation engine

The simulation engine is an optional engine necessary to perform the Monte-Carlo simulation presented throughout the various chapters of this book. A differentiation has been made between the standard and the advanced simulation engine, as follows:

- **Standard simulation:** The standard simulation engine allows a quick and easy simulation of the baseline schedule based on one of the 9 scenarios as shown in table 4.2.
- **Advanced simulation:** An advanced distribution drawer allows the user to generate random durations from a predefined distribution. Each project activity has a default triangular distribution with an average equal to its baseline duration. A distribution drawer can be used to manually change the default distribution to any distribution (on the project level for all activities or on the individual activity level) which can be saved as self-created distributions usable for later simulation studies.

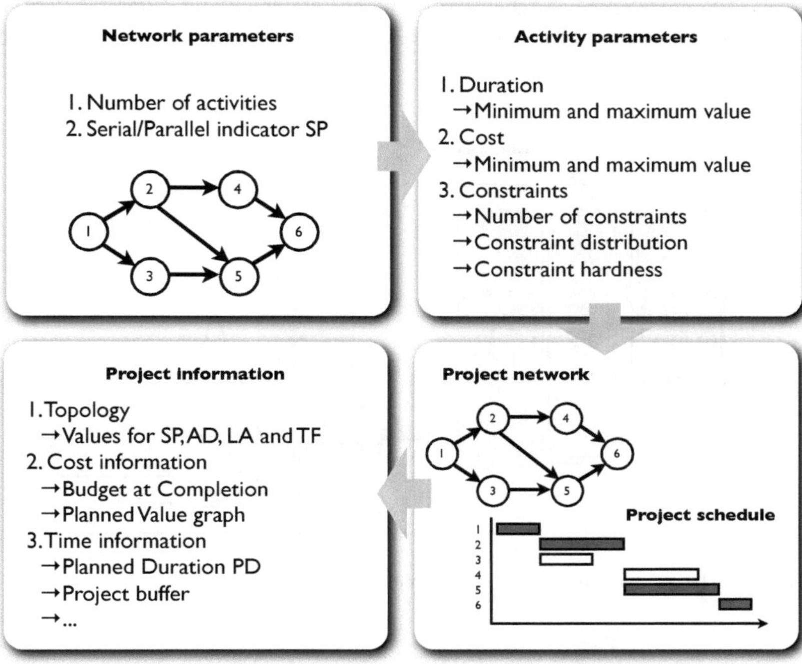

Fig. 7.7 The project network generation engine in ProTrack

The output of simulation runs obtained from both the standard and advanced simulation engine can be twofold depending on the version of ProTrack. First, the output can be detailed sensitivity information for each individual activity which serves as a schedule risk analysis (applicable in the "Sensitivity Scan" or "Smart Version" of ProTrack, see figure 7.9), as discussed in chapter 5. Second, the simulation engine can be used to generate a forecast accuracy study, which will be explained in the time forecasting engine of next section (applicable in the "Time Shuttle" or "Smart Version") and is similar to the study of chapter 4.

Figure 7.8 shows the standard and advanced simulation engine options of ProTrack. The standard simulation engine allows a quick and easy simulation study and corresponds to the 9 scenarios as shown in table 4.2. Recall that these scenarios contain both "normal" project behavior scenarios as well as more "extreme" project behavior scenarios, which make this simulation option especially useful for the EVM simulation and accuracy studies presented in chapter 4. The advanced simulation engine allows a more detailed and refined simulation study by assigning predefined distributions to each individual project activity. The advanced simulation engine is constructed such that both a more experienced user as well as a newcomer in the schedule risk analysis can quickly generate project sensitivity information. Each activity duration has by default a simple triangular distribution, but this can be changed to more advanced distribution by simple drags and drops. The sensitivity information obtained after a simulation run is similar to the study described in chapter 5.

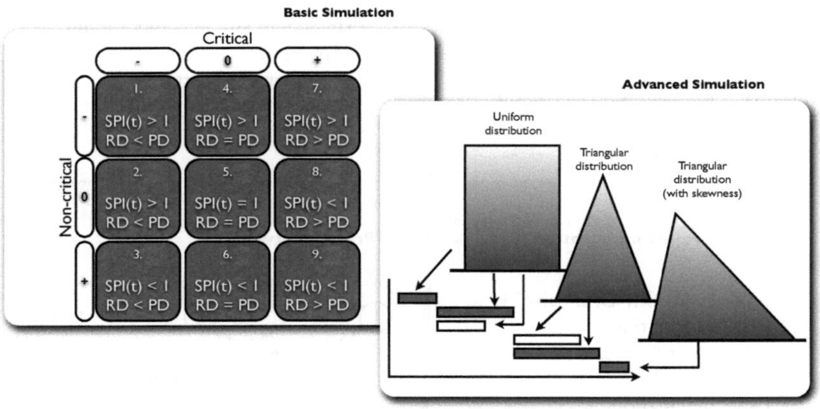

Fig. 7.8 The standard and advanced simulation engine options in ProTrack

7.3.3 Time forecasting engine

The time forecasting engine generates a single set or multiple sets of project tracking periods to measure fictitious earned value data and to report the corresponding forecast accuracy. The time forecasting engine can be used for two purposes, as follows:

- **Tracking period generator:** This part of the time forecasting engine can be considered as a fictitious tracking period generator where the active project is subject to a simulated tracking run until the project finish. The automatic creation of tracking periods as well as the ability to review the performance status of periods back in time allows to learn the behavior of the project over time and the evolution of earned value based metrics for the particular project under study. This engine option is particularly useful for newcomers to get acquainted with EVM and ES, or can also act as a research tool for EVM/ES researchers and practitioners.
- **Forecast accuracy calculator:** This version of the time forecasting engine allows multiple simulation runs which are required to perform a forecast accuracy experiment similar to chapter 4. Unlike the tracking period generator that generates a single set of tracking periods to forecast a single fictitious project execution, this calculator generates multiple sets of tracking periods (i.e. multiple

fictitious project executions). This calculator allows the user to learn and to get acquainted with the behavior of the active project under a wide variety of fictitious circumstances and scenarios. Consequently, this accuracy calculator makes use of the standard and/or advanced simulation engine of ProTrack and reports average forecast accuracy results by reporting values for the MAPE or MPE as shown in chapter 4.

7.3.4 4 versions of ProTrack

ProTrack is released under 4 different versions each with a specific functionality summarized in figure 7.9. The main functionalities basically differ in the engines incorporated in ProTrack and can be briefly described along the following lines[2]:

- **Standard Version:** ProTrack standard contains the standard CPM scheduling module and the EVM tracking module (but no engines), including the calculation of the earned schedule metric and the time forecasts for the three forecasting methods ($EAC(t)_{PV}$, $EAC(t)_{ED}$ and $EAC(t)_{ES}$).
- **Sensitivity Scan:** ProTrack standard version extended with the project generation engine and the advanced simulation engine. Hence, this version allows the automatic generation of project network data as well as multiple advanced simulation runs from predefined distributions to scan the sensitivity of all project activities, as discussed in chapter 5.
- **Time Shuttle:** ProTrack standard version extended with the project generation engine, the standard simulation engine and the time forecasting engine. Consequently, this ProTrack version allows the generation of fictitious tracking periods from the start till the finish of an existing or automatically created project network which automatically generates earned value based tracking information. Moreover, multiple simulation runs based on one of the 9 scenarios to measure the forecast accuracy of the known EVM methods also belong to the possibilities of this ProTrack version.
- **Smart Version:** ProTrack smart version contains all engines as well as a copy of this book "Measuring Time - An Earned Value Simulation Study" and is the most complete version that allows both practitioners as well as academics to learn how EVM works in theory and practice. Both the simulation engine and the time forecasting engine can be used under the standard and advanced simulation module, and the ability to measure the forecast accuracy of earned value based methods discussed throughout the previous chapters can be tested for a single run tracking period generation or for multiple simulation runs.

---

[2] ProTrack has been released end of 2008 and will be subject to a never-ending stream of improvements, changes and state-of-the-art updates taking wishes and needs of OR-AS customers into account. For an updated list of ProTrack functionalities, check www.or-as.be/protrack.

CRITERIA	ProTrack Standard Version	ProTrack Sensitivity Scan	ProTrack Time Shuttle	ProTrack Smart Version
Project Planning				
CPM Planning	yes	yes	yes	yes
Network Visualization	-	yes	yes	yes
Automatic Project Creation	-	yes	yes	yes
Project Simulation Options				
Activity Time/Cost Simulation	-	Advanced Module	Standard Module	Standard/Advanced Modules
Schedule Risk Analysis	-	yes	-	yes
EVM Forecast Accuracy Analysis	-	-	yes	yes
Project Tracking				
Earned Value and Earned Schedule	yes	yes	yes	yes
Earned Value Forecasting	yes	yes	yes	yes
Automatic Tracking Generation	-	-	yes	yes
p-factor Calculation	-	-	yes	yes
Book "Measuring Time"	-	-	-	yes
Price	299 €	399 €	399 €	599 €

Fig. 7.9 The 4 ProTrack options: Standard Version - Sensitivity Scan - Time Shuttle - Smart Version

7.4 Demo experiment

The purpose of this section is to present a short review of the studies presented in the previous chapters of this book by using ProTrack on 9 projects generated by the project generation engine. The intention is to illustrate case-specific results in order to confirm the results found throughout the various chapters, without re-showing and generalizing results. The illustration is done based on the project life cycle, and searches for static and dynamic determinants of earned value based forecast accuracy.

Typically, a project goes through a number of different phases, which is often referred to as the project life cycle. This cycle has been described extensively by many authors (see e.g. the Project Management Body of Knowledge (PMBOK, 2004)) and consists of a project conception phase, a project definition phase, a phase in which the project has to be scheduled, the execution of the project, the project control phase which monitors the current performance of the project and the termination and/or evaluation of the project. Figure 7.10 displays a 6 phased example project life cycle that will be used throughout this chapter.

The demo experiment deals with the control phase of the project life cycle, and the corresponding performance measurement feedback loop (see figure 7.10) from project control to the planning/scheduling phase. More precisely, the focus is on a reactive scheduling and performance measurement system using earned value management (EVM). The aim is to test the accuracy of various earned value predictive methods to measure and forecast the final duration (i.e. time focus, not cost focus) of a project and to search for determinants that affect the accuracy of these predictive methods. In a way, this is a simple repetition of the studies presented in this book.

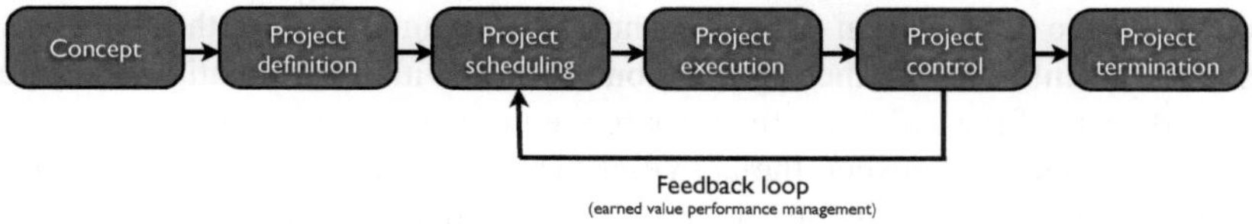

Fig. 7.10 The project life cycle (PLC) with a reactive scheduling approach

Section 7.4.1 presents four hypotheses in order to search for static (i.e. before the start of the project) as well as dynamic (i.e. during project execution) determinants that affect the forecast accuracy of the predictions. These hypotheses have been defined based on the results of the previous chapters, and consequently, serve as illustrations rather than new research hypotheses. Section 7.4.2 makes use of ProTrack to test and confirm or reject the hypotheses based on the analysis of the 9 generated projects. Each of the projects will be simulated under various settings.

7.4.1 Determinants of forecast accuracy

In this section, the simulation results are analyzed in search of determinant factors that influence the accuracy of earned value based predictive methods to forecast a project's final duration. A distinction has been made between static determinants, which can be calculated before the start of the project (i.e. during definition and scheduling phase) and dynamic determinants, which can be calculated during the project's execution phase. Figure 7.11 gives an overview of the four determinants along the life of the project that will be discussed in the next subsections.

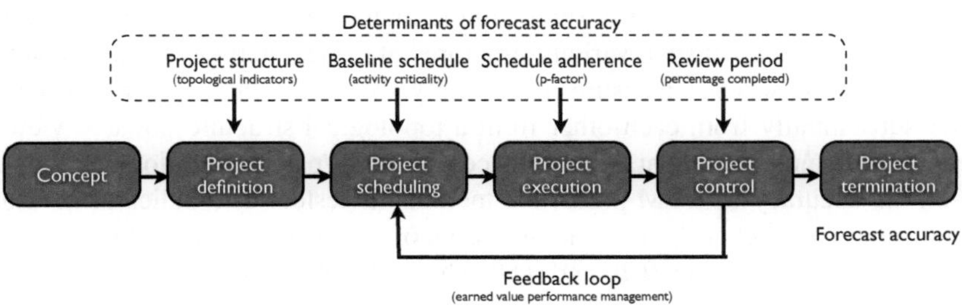

Fig. 7.11 Static and dynamic determinants of EVM accuracy

At the beginning, in the so-called conceptual phase, an organization identifies the need for a project or receives a request from a customer.

In the definition phase the project manager defines the project objectives, the project specifications and requirements and the organization of the whole project. In doing so, the organization decides on how it is going to achieve these objectives. Based on this information, the organization can start with the estimation of the durations and costs of the activities, the resource requirements and availabilities and the precedence relations between the activities. The next step is the scheduling phase in order to present a timetable for the project activities resulting in a (resource-feasible) baseline schedule. In the next subsection, static determinants (referring to periods before the start of the project, i.e. the definition and scheduling phase) will be discussed to search for factors that influence the predictive power of earned based methods to forecast a project's final duration during the execution phase.

During the execution phase and the control phase the project has to be executed, monitored and controlled to detect whether it is performed according to the existing schedule. If deviations occur, corrective actions have to be taken which results in a reactive feedback loop to bring the project back on track. A second subsection refers to dynamic determinants during the life of the project that affect the quality of the earned value based predictive methods.

The termination phase involves the completion and a critical evaluation of the project. This information can then be used during the project life cycle of future, similar projects since the specifications of a project, the estimates of the durations and costs and many things more are often determined on the basis of averages of past performance.

7.4.1.1 Static determinants of forecast accuracy

The static determinants during the preparation phases of the project (i.e. definition and scheduling) have been displayed in figure 7.12, and will be summarized along the following lines of the section.

Definition phase: The topological structure of a project network, defined by the number and distribution of the activities and their precedence relations, can be easily measured through the use of often simple mathematical calculations of indicators that distinguish between various structures of project networks (see chapter 4). These indicators serve as measures of diversity able to detect project networks that differ substantially from each other from a topological structure point of view. In the current demo experiment, the influence of a serial/parallel topological indicator SP on the accuracy of EVM predictive methods is tested. The indicator measures the closeness of a network to a complete serial or complete parallel network and has been set to values between 0.20 and 0.80. The first hypothesis can be formulated as follows:

H1: The topological structure of a project network has an influence on the accuracy of EVM duration forecast measures.

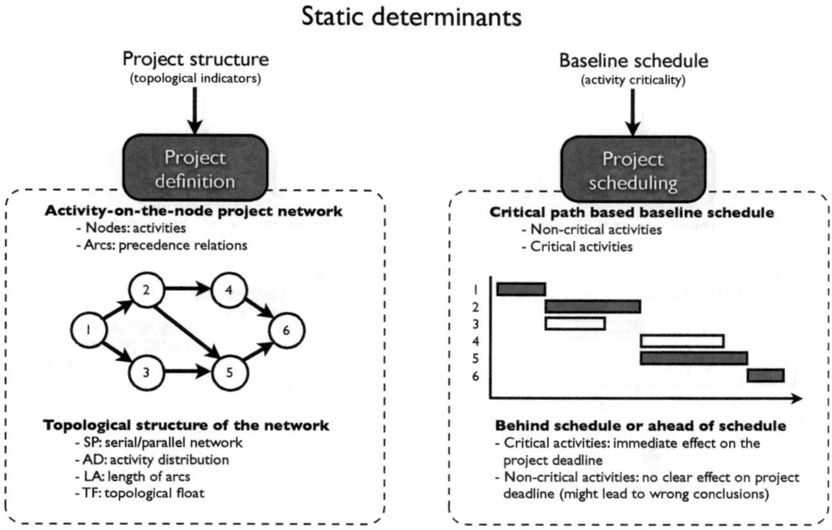

Fig. 7.12 Static determinants of EVM accuracy during project definition and scheduling phase

Scheduling phase: Project scheduling aims at the construction of a baseline schedule where each activity is sequenced subject to the precedence constraints (known as the critical path method) and possible resource constraints (known as the resource-constrained critical chain method). Traditional forward pass calculations result in the presence of a critical path (without resources) or a critical chain (with resources). In chapter 4, it has been shown that the activity criticality heavily determines the accuracy of earned value based metrics, since changes (delays or accelerations) in critical activities have an immediate effect on the project duration, while changes in non-critical activities might have no effect at all on the final duration of the project. The second hypothesis can be formulated as follows:

H2: The activity criticality reported by a project's baseline schedule has an influence on the accuracy of EVM duration forecast measures.

7.4.1.2 Dynamic determinants of forecast accuracy

Figure 7.13 gives an overview of the dynamic determinants of EVM forecast accuracy during the life of the project. The adherence of the original baseline schedule (execution phase) as well as the choice of the length of the review period between start and end of the project provide dynamic information about the accuracy of the project schedule performance. Details of these two sources of dynamic information parameters are described along the following lines.

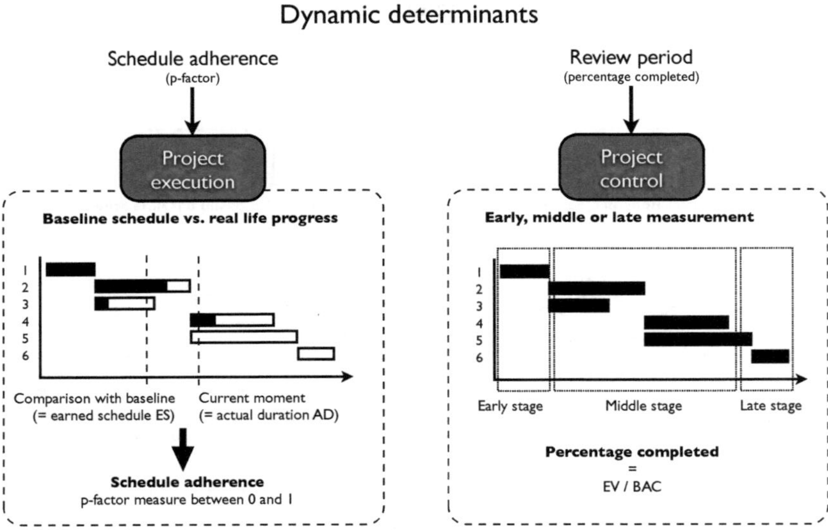

Fig. 7.13 Dynamic determinants of EVM accuracy during project execution and control phase

Execution phase: During the execution of the project, the original activity timetable can be disrupted due to numerous reasons leading to a project execution that is not in congruence with the original baseline schedule. This lack of schedule adherence can be dynamically measured through the use of the p-factor initially proposed by Lipke (2004) and presented in chapter 2. The p-factor value is a simple measure based on the calculation of the traditional EVM metrics (AC, PV and EV) and measures the portion of the work that is done in congruence with the original baseline schedule. While the ES metric measures the current duration performance compared to the baseline schedule and indicates whether the project is ahead or behind schedule, the p-factor measures the performance of the project relative to this ES metric, and hence, measures the degree of schedule adherence given its current (good or bad) performance.

The lack of schedule adherence can be contributed to numerous factors, among which activity overlapping (out-of-sequence execution of activities), deviations from the original baseline activity estimates (in terms of time and cost) or non-linear EV accrue have been tested in section 4.3. The third hypothesis tested in this chapter can be described as follows:

H3: The degree of schedule adherence has an influence on the accuracy of EVM duration forecast measures.

Control phase: During the control phase, the decision maker (i.e. the project manager) has to determine the length of the review periods as well as the interval in which EVM based predictive metrics might produce reliable results. A crucial assumption of EVM based forecasting is that the prediction of the future is based on information on performance from the past (see table 4.2), and hence, unreliable data from the past might give false predictions to the future. It is therefore of crucial importance to determine the time window in which the EVM metrics produce more or less reliable results. Undoubtedly, the accuracy of forecasts depends on the completion stage of the project. Obviously, the EVM metrics measured at the very beginning of the project are often very unreliable due to the lack of sufficient and adequate data to assume that future performance will follow the current performance. Moreover, the classic schedule indicators (SV and SPI) have been shown to be unreliable as project duration forecasting indicators since they show a strange and unreliable behavior over the final third of the project which has led to the development of the earned schedule concept by Lipke (2003) which behaves correctly over the complete project horizon.

For these reasons, the accuracy of index-based time forecasts is measured as a function of the completion stage of the project, measured by the percentage completed EV / BAC, and determines the average accuracy of the predictions made in the early, middle and late stages of the project execution phase, both for projects ahead of schedule or for projects with a delay. A last hypothesis can be written as follows:

H4: The earned schedule method is more reliable than the traditional earned value methods in all stages of the project.

7.4.2 ProTrack simulation experiment

The four hypotheses have been tested on a small set of fictitious projects that have been generated by the generation engine of ProTrack, each containing 100 activities. The average project baseline duration is somewhat more than a year and the tracking period is set to one week.

Table 7.2 9 projects used in the ProTrack demo experiment

ID	1	2	3	4	5	6	7	8	9
PD (in weeks)	36	42	45	51	55	59	68	72	85
SP	0.20	0.25	0.30	0.45	0.50	0.55	0.70	0.75	0.80

Table 7.2 displays the characteristics of the 9 projects used for the demo experiment. In the remainder of this section, brief results will be discussed.

7.4.2.1 Hypothesis 1. Topological network structure

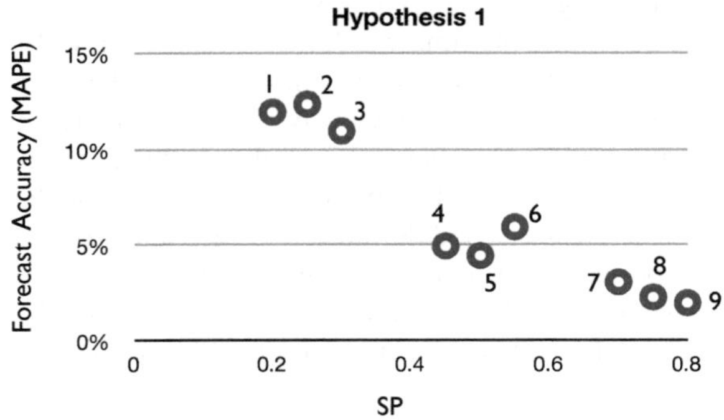

Fig. 7.14 Demo results for hypothesis 1

Figure 7.14 displays the $EAC(t)_{ES2}$ accuracy (measured by the MAPE) for the 9 projects and shows that the topological structure of a project network has a clear influence on the accuracy of EVM duration forecast measures. The more the project resembles a serial network (high SP), the higher the accuracy (low MAPE) of the predictive measures. The figure confirms hypothesis 1 and the results found in chapter 4.

7.4.2.2 Hypothesis 2. Activity criticality

In order to test the validity of hypothesis 2, the criticality index CI has been calculated for each activity, and the average CI value is reported for each project network.

Figure 7.15 shows the $EAC(t)_{ES2}$ accuracy results for the 9 projects and clearly confirms the second hypothesis and shows that the average activity criticality has a clear influence on the accuracy of the predictive methods as follows: the higher the activity criticality, the better the accuracy of the forecasts. Obviously, a lower activity criticality means a lower probability of being on the critical path, and hence, the more likely a delay (within the activity slack) reported by the SPI or SPI(t) indicators, has no effect on the final project deadline. Consequently, the number of critical activities observed in the baseline schedule gives a first rough indication on the accuracy of the future EVM tracking performance measures (i.e. confirmation of hypothesis 2).

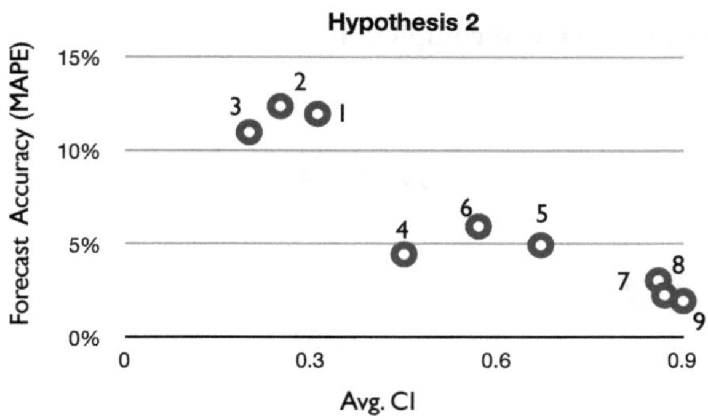

Fig. 7.15 Demo results for hypothesis 2

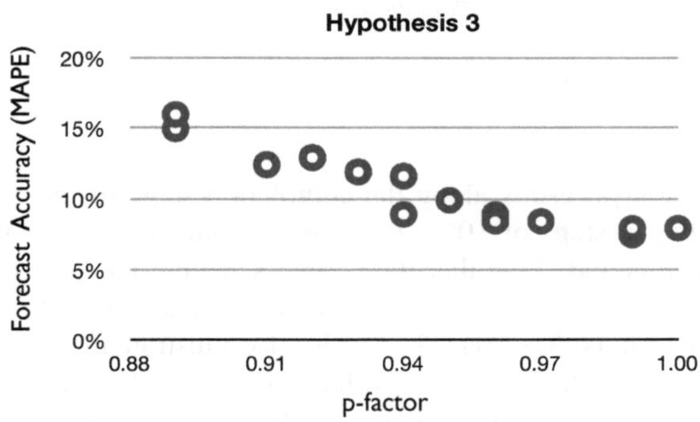

Fig. 7.16 Demo results for hypothesis 3

7.4.2.3 Hypothesis 3. Schedule adherence

Figure 7.16 shows the relation between the average p-factor and the average $EAC(t)_{ES2}$ forecast accuracy for a subset of the 9 projects simulated under 15 different scenarios (with and without overlapping, linear or non-linear EV/PV accrue, etc. (For more information, see chapter 4). The scatterplot shows a negative relation between the average p-factor and the forecast accuracy, i.e. lower p-factor values denoting a certain lack of schedule adherence often result in less accurate forecasts (i.e. confirmation of hypothesis 3). Hence, the p-factor, which can be dynamically measured during a review of the project (i.e. project tracking) based on the traditional EVM metrics, can be considered as a dynamic warning signal of the duration forecast accuracy.

7.4.2.4 Hypothesis 4. Stage of completion

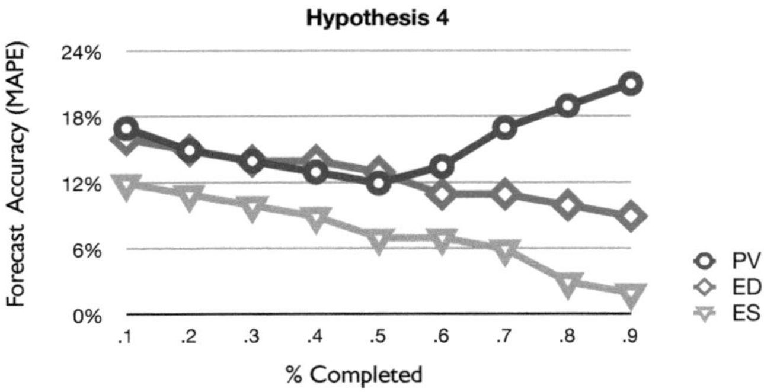

Fig. 7.17 Demo results for hypothesis 4

Figure 7.17 displays results along the completion stage of the project, varying from 10% to 90%, in steps of 10%. The results from the figure confirm the previously found results and show that the earned schedule method outperforms, on average, the other forecasting methods in all stages of the project execution phase (i.e. confirmation of hypothesis 4). The results also illustrate the quirky behavior of the SPI indicator (used in the planned value and earned duration methods) at the late stage of the project. Indeed, the late stage forecast accuracy is much better for the ES method compared to the PV and ED methods. The results for the planned value method show that the use of the SPI indicator, which goes to a final value of 100%, regardless of the project performance, leads to very low quality predictions at the late stage of the project. The SPI(t) indicator of the earned schedule method has been developed to overcome this quirky behavior, leading to an improved forecast accuracy at the end of the project. Obviously, measuring project performance and predicting future performance based on the resulting data leads to the lowest accuracy at the early stages of the project execution phase.

7.5 Conclusion

In this chapter, the main features of ProTrack have been reviewed in detail and a demo experiment has been shown to briefly illustrate the link with the research study of this book.

ProTrack (acronym for Project Tracking software) is a novel software tool launched on the market mid 2008 that combines project scheduling with earned value simulation and forecast accuracy measurement. The software combines project scheduling and tracking with earned value management and allows the calculation of a project's time forecast using the three EVM methods extensively discussed in this book. On top of that, a project generation, a (standard and advanced) simulation and a time forecasting engine can be added to randomly generate project networks and simulate the execution phase in order to test a baseline schedule's sensitivity (this is known as schedule risk analysis, see Hulett (1996)) as well as to measure the accuracy of EVM based predictive methods. The project tracking steps are automatically saved along the life of the project, allowing to measure the accuracy of the predictions and to review the current status of the project at any moment back in time.

The four hypotheses discussed in this chapter are illustrations of the studies performed in the previous chapters of this book. The project generation engine allows the creation of fictitious projects with a controlled structure (hypothesis 1). A simulation engine enables the user to measure the criticality of individual project activities (hypothesis 2). The earned value/earned schedule module calculates the more traditional EVM methods as well as the earned schedule metrics including the p-factor (hypothesis 3). The simulation and time forecasting engines allow the simulation of activity duration variations and measure the accuracy of project duration forecasts along various stages of the project life cycle (hypothesis 4). Details can be found on www.or-as.be/protrack.

Chapter 8
Conclusions

Earned value systems have been set up to deal with the complex task of controlling and adjusting the baseline project schedule during execution, taking into account project scope, timed delivery and total project budget. It is a well-known and generally accepted management system that integrates cost, schedule and technical performance. It is mainly used to calculate cost and schedule variances, performance indices and forecasts of a project's final cost and duration. The earned value method provides early indications of project performance to highlight the need for eventual corrective actions.

The research study of this book deals with the project performance and control phase of the project life cycle, and the corresponding feedback loop from control to planning and scheduling to take corrective actions when necessary. More precisely, the focus is on a reactive scheduling early warning system by means of earned value metrics. Although EVM has been set up to follow up both time and cost, the majority of the research has been focused on the cost aspect. Recently, different sources in literature show that the "classic" earned value metrics fail in predicting the total project duration in an accurate way. The research study in this book aims at filling that gap and investigates the time dimension of earned value management systems and their potential to predict the final duration of a project.

The *planned value method* and the *earned duration method* are two well-known methods that rely on the traditional schedule performance indicator SPI and can be used to predict a project's final duration. The *earned schedule method* has been developed as a criticism on the use of the classic SV and SPI metrics since they give false and unreliable time forecasts near the end of the project. Instead, two alternative schedule performance measures (referred to as SV(t) and SPI(t)) that are directly expressed in time units have been presented to overcome the quirky behavior of the classic SV and SPI indicators.

Since its introduction, the earned schedule method has been investigated both from a theoretical point of view as from a practitioner's point of view. Empirical evidence has been provided on a few real life projects in the United States, the United Kingdom, Australia and Belgium, amongst some others. A theoretical extension of the earned schedule method, known as the p-factor, has been recently proposed in

M. Vanhoucke, *Measuring Time*, International Series in Operations Research & Management Science 136, DOI: 10.1007/978-1-4419-1014-1_8,

order to dynamically measure the schedule adherence of a project during its execution. The measure is a straightforward derivation of the simple ES formula and allows the calculation of project impediments or constraints. This novel concept gradually finds its way to real life projects and add-ins to existing software tools, but limited empirical evidence is available up to today.

The limited empirical evidence of the earned value based time predictive methods has led to the case study of chapter 3. Together with some efforts done in the US, the UK and Australia, Fabricom Airport Systems (currently known as Logan Teleflex) can be considered as an early adopter of the earned schedule technique. Although the real life study at Fabricom has its merits, and has contributed to the theoretical development of the research study of this book, results obtained by a case study are often too case-specific which makes it difficult to draw general conclusions due to the small sample of data (only full data for three projects were available). The simulation studies in chapters 4 and 5 aimed at generating more general results.

8.1 Forecast accuracy

In the simulation studies of chapter 4, the forecast accuracy of the three methods has been validated by simulations on a large and diverse set of projects under various controlled scenarios.

The results show that under "normal" circumstances the earned schedule method has the best performance, leading to small deviations between the duration forecast and the final project duration. Normal circumstances are defined as project progress where the schedule performance indicators report reliable results during the life of the project. However, special scenarios have been simulated to force the schedule performance indicators to report unreliable results. Under these "extreme" circumstances, the earned schedule method performs worse than the earned duration and planned value methods. Consequently, the earned schedule method can be considered as a reliable time forecasting method, as the method's forecast is strongly based on the quality of the schedule performance indicator value (SPI(t)), and is able to forecast the final project duration in an accurate way when the schedule performance indicator SPI(t) reports a correct warning signal about the current project performance.

The simulation studies have also revealed that the topological network structure has a clear and strong influence on the time forecast accuracy of the various methods. More precisely, an indicator that measures the closeness of a project network to a complete parallel or serial network has been used throughout the study, and has shown that the time predictions are relatively more accurate for projects with a lot of serial activities compared to more parallel project networks.

8.2 Schedule adherence

In a second simulation study of chapter 4, a dynamic schedule adherence concept, known as the p-factor approach, has been embedded in the simulation runs to test its ability to dynamically predict and improve the forecast accuracy.

Results have shown the evolution of schedule adherence as an improving measure always ending at 100% at the end of the project, and have shown a relation between the schedule adherence and forecast accuracy of the three predictive methods. The main contribution of the p-factor lies in the ability to calculate the effective earned value to detect project impediments and/or constraints by taking the risk of rework into account. However, the effective earned value concept is not able to establish accuracy improvements in earned value predictions.

Due to the limited empirical evidence available to support the conclusions made in this chapter and the limited contribution of the p-factor approach to improve the accuracy of the time forecasts, the results obtained in this chapter should be considered as very preliminary and more research is necessary. However, the schedule adherence concept certainly acts as an interesting eye-opener to the need of a more dynamic measure to calculate and improve the forecast accuracy. The concept has certainly contributed to the renewed attention to earned value based time forecasting research, and will hopefully stimulate both academics and practitioners to continue their current research efforts and set up new test experiments to investigate the contribution of the p-factor concept to project performance measurement. To be continued, hopefully...

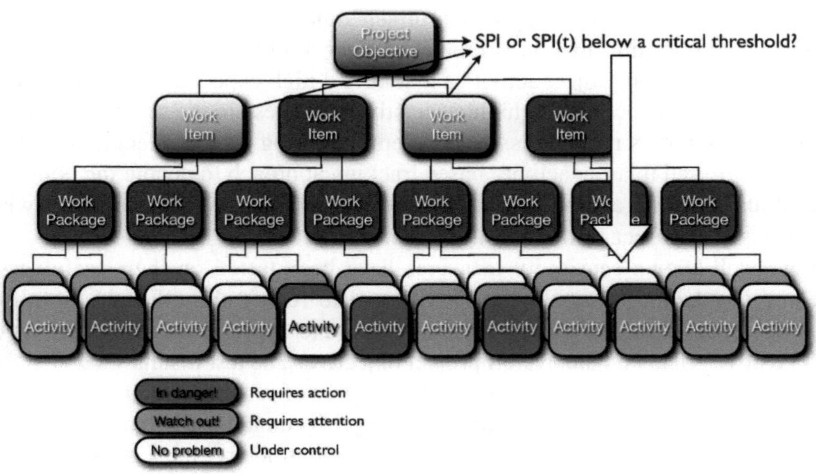

Fig. 8.1 The top-down project based tracking approach of earned value management

Throughout the various chapters of this book, it has been noted that project tracking using earned value management should not be considered as an alternative to the well-known critical path based scheduling and tracking tools. Instead, the EVM methodology offers the project manager a tool to calculate a quick and easy sanity check on the control account level or even higher levels of the work breakdown structure (WBS). In this respect, an earned value management system is set-up as an early warning signal system to detect problems and/or opportunities in an easy and efficient way, which is obviously less accurate than the detailed critical path based scheduling analysis of each individual activity. However, this early warning signal, if analyzed properly, defines the need to eventually drill down into lower WBS levels. In conjunction with the project schedule, it allows taking corrective actions on those activities which are in trouble (especially those tasks which are on the critical path). In this book, this top-down tracking approach is called a project based tracking method. Figure 8.1 displays a fictitious work breakdown structure (WBS) to illustrate the project based project tracking approach of earned value management.

8.3 Time sensitivity

The simulation study of chapter 5 has been set up as a reaction to the poor forecast accuracy of EVM predictive methods when the project is more parallel. The goal of this study is to investigate whether activity sensitivity information can be used to guide the project tracking process as an alternative for the weak accuracy for the time forecasting methods on project networks with a lot of parallel activities. Figure 8.2 illustrates the bottom-up approach of schedule risk analysis. The detection of activity sensitivity information is crucial to steer a project manager's attention towards a subset of the project activities that have a high expected effect on the overall project performance. These highly sensitive activities are the subject to intensive control, while others require less or no attention during project execution. This approach is referred to as an activity based tracking approach to denote the bottom-up control and tracking approach to take corrective actions on those activities with a highly expected effect on the overall project objective.

Four well-known sensitivity measures have been tested on their usefulness to measure the degree of activity sensitivity and to reduce the effort of the project tracking process without losing the ability to take appropriate corrective actions with positive effects on the overall project objective. The test results show that most sensitivity measures are able to measure the degree of sensitivity and can be used as identifiers of an activity's sensitivity when projects contain many parallel activities. However, for projects with a more serial network structure, most sensitivity measures are no longer able to distinguish between insensitive and sensitive activities, and hence, a careful selection of a subpart of the activity set that will be subject to a detailed tracking approach is more difficult or simply impossible. The overall conclusion is that the criticality index CI, the significance index SI and the cruciality index CRI perform well for parallel networks but fail in discriminating between low and high sensitivity for serial networks. The schedule sensitivity index is the only sensitivity measure that is able to select a sensitive subset of activities for both parallel and serial networks, and hence, can be easily used to guide and simplify the bottom-up tracking process.

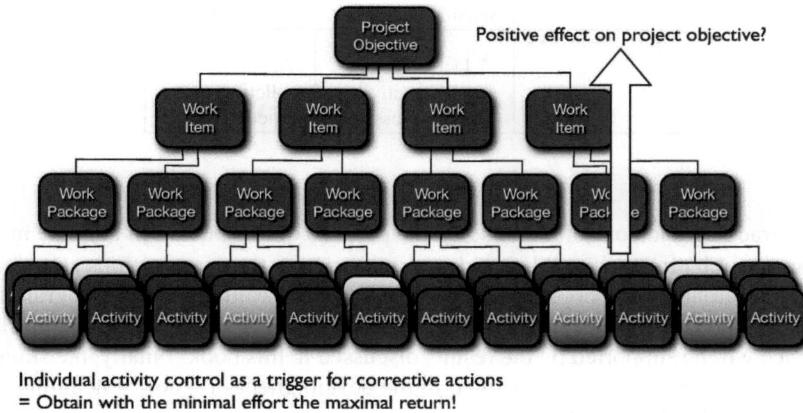

Fig. 8.2 The bottom-up activity based tracking approach of schedule risk analysis

8.4 Summary

Chapter 6 has experimentally validated the efficiency of the two alternative project tracking methods of figures 8.1 and 8.2 by means of a fourth simulation study. Table 8.1 summarizes the main conclusions of the four simulation studies presented throughout this book. The simulation studies of chapter 4 have clearly demonstrated that a top-down project based tracking approach using the earned duration or earned schedule methods provides highly accurate results when the project network contains more serial activities. This top-down approach lies in the heart of the earned value management philosophy and has been tested in detail throughout this book. The bottom-up activity based tracking approach using sensitivity information of activities obtained through a standard schedule risk analysis is particularly useful when projects contain a lot of parallel activities. This bottom-up approach requires often subjective distribution information of individual activities which implies a certain activity risk estimate, but simplifies the tracking effort to those activities with a high expected effect on the overall project objective.

I sincerely hope that this book acts as a summary and overview of the often confusing and case-specific research results spread throughout the more popular literature, and keeps stimulating the future research efforts in the domain of project moni-

Table 8.1 Overall summary of simulation studies

	Activity based project tracking (bottom-up)	Project based project tracking (top-down)
Parallel networks	√ Focus only on highly sensitive activities	X Inaccurate time predictions
Serial networks	X Detection of sensitive activities often impossible	√ Accurate time predictions (using earned schedule)

toring, tracking and control. Personally, the purpose of this book will help me in my future research efforts. It will be used as a guidance for the presentations at national and international workshops, the in-company earned value teaching programmes and the consultancy projects. Moreover, the ever on-going search to more empirical evidence will be supported by the results discussed in this book. Finally, the simulation studies of this book have resulted in the development of a new scheduling and tracking tool ProTrack. After years of study from both a theoretical and practical point of view, I believe ProTrack contains all necessary features such that the user can repeat almost all experiments written and discussed in this book. Of course, the purpose of ProTrack is more than simply a research tool accompanied by a book. ProTrack offers a traditional software scheduling and project tracking tool as an alternative to the many tools available on the market. However, the options to perform several kinds of expert analyses is, to the best of my knowledge, unique in its kind. The combination of the research results written in this book and the expert engines available in ProTrack allows the user to learn the do's and do not's of project scheduling, tracking and risk analysis, and will hopefully reduce the black box problem of most project scheduling software.

The research study can be useful for project managers for small and medium sized enterprises and aims to give (partial) answers on the following questions:

- Which method do you use best for your project with given characteristics?
- What is the expected accuracy of your project measurement system?
- What are the parameters that influence your project performance measurement accuracy?
- When do you take corrective actions (= project tracking)?

Consequently, I believe that the relevance of the research written in this book might be substantial to both academics as well as practitioners for three reasons. First, this is, to the best of my knowledge, the first research study that evaluates the three earned value methods as predictors of a project's final duration in a profound way. Until now, the research has been limited to occasional case-studies published in non-peer reviewed journals. However, both academic people and practitioners need a profound and detailed comparison of the methods in order to gain understanding in the behavior of the methods. Although many research has been devoted to the cost-related earned value metrics (published in peer-reviewed academic journals as well as in more popular magazines), I believe this book is the first summary that compares and validates the time-related earned value metrics. Second, although the research primarily focuses on a theoretical summary of EVM time predictions and an academic contribution to earned value/earned schedule management, the results and ideas are clearly inspired by many discussions with practitioners. Although many of the concepts discussed are only validated on a small set of real life projects, the practical validation is an on-going process and more empirical evidence is certainly on its way. Finally, the research study combines various other research efforts, published in flagship Operations Research journals such as the Journal of Scheduling, the European Journal of Operational Research, the Journal of the Operational Research Society and many more.

Obviously, the research study presented in this book is only a first step in a long-term research goal. The research study of this book mainly aims at a validation and a detailed analysis of the three earned value based methods as alternative forecasting methods to predict a project's final duration. Future research will undoubtedly be necessary in order to improve the understanding of a project's performance behavior and the search to drivers of project performance in reality. I truly hope that this book can act as a stimulator of the necessary further research and will finally contribute to an improved project performance measurement.

As a final but important note: the research presented in this book was not possible without the financial support of various sources. The support by the Research collaboration fund of PMI Belgium (2007), the research project funding by the Flemish Government (2008) and the research project under the contract name G.0194.06 of the FWO (2005-2009) is acknowledged and greatly appreciated.

References

Agrawal M, Elmaghraby S, Herroelen W (1996) DAGEN: A generator of testsets for project activity nets. European Journal of Operational Research 90:376–382
Akkan C, Drexl A, Kimms A (2005) Network decomposition-based benchmark results for the discrete time-cost tradeoff problem. European Journal of Operational Research 165:339–358
Alvarez-Valdes R, Tamarit J (1989) Heuristic algorithms for resource-constrained project scheduling: a review and empirical analysis. In Slowinski, R. and Weglarz, J. (eds.), Advances in Project Scheduling. Elsevier, Amsterdam
Amor J (2002) Scheduling programs with repetitive projects using composite learning curve approximations. Project Management Journal 33:16–29
Amor J, Teplitz C (1993) Improving CPM's accuracy using learning curves. Project Management Journal 24:15–19
Amor J, Teplitz C (1998) An efficient approximation procedure for project composite learning curves. Project Management Journal 29:28–42
Anbari F (2003) Earned value project management method and extensions. Project Management Journal 34:12–23
Badiru A (1995) Incorporating learning curve effects into critical resource diagramming. Project Management Journal 2:38–45
Bein W, Kamburowski J, Stallmann M (1992) Optimal reduction of two-terminal directed acyclic graphs. Siam Journal on Computing 21:1112–1129
Book S (2006a) Correction note: "Earned Schedule" and its possible unreliability as an indicator. The Measurable News Fall:22–24
Book S (2006b) "earned schedule" and its possible unreliability as an indicator. The Measurable News Spring:24–30
Cho J, Yum B (1997) An uncertainty importance measure of activities in PERT networks. International Journal of Production Research 35:2737–2758
Christensen D (1993) The estimate at completion problem: A review of three studies. Project Management Journal 24:37–42
Cooper K (2003) Your project's real price tag? Letters to the editor. Harvard Business Review 81:122–122

Covach J, Haydon J, Riether R (1981) A study to determine indicators and methods to compute estimate at completion (EAC). Virginia: ManTech international corporation.

Dar-El E (1973) MALB - A heuristic technique for balancing large single-model assembly lines. IIE Transactions 5:343–356

Davies E (1974) An experimental investigation of resource allocation in multiactivity projects. Operational Research Quarterly 24:587–591

Davis E (1975) Project network summary measures constrained-resource scheduling. AIIE Transactions 7

De Reyck B (1995) On the use of the restrictiveness as a measure of complexity for resource-constrained project scheduling. Research Report 9535 Department of Applied Economics, Katholieke Universiteit Leuven, Belgium

De Reyck B, Herroelen W (1996) On the use of the complexity index as a measure of complexity in activity networks. European Journal of Operational Research 91:347–366

Demeulemeester E, Dodin B, Herroelen W (1993) A random activity network generator. Operations Research 41:972–980

Demeulemeester E, Vanhoucke M, Herroelen W (2003) A random network generator for activity-on-the-node networks. Journal of Scheduling 6:13–34

Dodin B, Elmaghraby S (1985) Approximating the criticality indices of the activities in PERT networks. Management Science 31:207–223

Drexl A, Nissen R, Patterson J, Salewski F (2000) ProGen/πx - an instance generator for resource-constrained project scheduling problems with partially renewable resources and further extensions. European Journal of Operational Research 125:59–72

Elmaghraby S (1977) Activity networks: Project planning and control by network models. New York: John Wiley and Sons, Inc.

Elmaghraby S (2000) On criticality and sensitivity in activity networks. European Journal of Operational Research 127:220–238

Elmaghraby S, Herroelen W (1980) On the measurement of complexity in activity networks. European Journal of Operational Research 5:223–234

Elmaghraby S, Kamburowsky R (1992) The analysis of activity networks under generalized precedence relations. Management Science 38:1245–1263

Elmaghraby S, Fathi Y, Taner M (1999) On the sensitivity of project variability to activity mean duration. International Journal of Production Economics 62:219–232

Fatemi Ghomi S, Teimouri E (2002) Path critical index and activity critical index in PERT networks. European Journal of Operational Research 141:147–152

Fleming Q, Koppelman J (2003) What's your project's real price tag? Harvard Business Review 81:20–21

Fleming Q, Koppelman J (2004) If EVM is so good... why isn't it used on all projects? The Measurable News Spring:6–9

Fleming Q, Koppelman J (2005) Earned value project management. 3rd Edition. Newtowns Square, PA: Project Management Institute

Ford D, Sterman J (2003) Overcoming the 90% syndrome: Iteration management in concurrent development projects. Concurrent Engineering Research and Applications 11:177–186

Ford J, Fulkerson D (1962) Flows in networks. Princeton University Press, Princeton, New Jersey.

Gutierrez G, Kouvelis P (1991) Parkinson's law and its implications for project management. Management Science 37:990–1001

Gutierrez G, Paul A (2000) Analysis of the effects of uncertainty, risk-pooling, and subcontracting mechanisms on project performance. Operations Research 48:927–938

Hecht L (2007) Case study of earned schedule to do predictions. The Measurable News Winter 2007-2008:16–18

Henderson K (2003) Earned schedule: A breakthrough extension to earned value theory? A retrospective analysis of real project data. The Measurable News Summer:13–17, 21–23

Henderson K (2004) Further developments in earned schedule. The Measurable News Spring:15–17, 20–22

Henderson K (2005) Earned schedule in action. The Measurable News Spring:23–28, 30

Henderson K, Zwikael O (2008) Does project performance stability exist? A re-examination of CPI and evaluation of SPI(t) stability? Crosstalk - The Journal of Defense Software Engineering 21:7–13

Herroelen W, De Reyck B (1999) Phase transitions in project scheduling. Journal of the Operational Research Society 50:148–156

Hulett D (1996) Schedule risk analysis simplified. Project Management Network July

Icmeli-Tukel O, Rom W (1997) Ensuring quality in resource constrained project scheduling. European Journal of Operational Research 103:483–496

Jacob D (2003) Forecasting project schedule completion with earned value metrics. The Measurable News March:1, 7–9

Jacob D (2006) Is "earned schedule" an unreliable indicator? The Measurable News Fall:15–21

Jacob D, Kane M (2004) Forecasting schedule completion using earned value metrics? Revisited. The Measurable News Summer:1, 11–17

Kaimann R (1974) Coefficient of network complexity. Management Science 21:172–177

Kaimann R (1975) Coefficient of network complexity: Erratum. Management Science 21:1211–1212

Kao E, Queyranne M (1982) On dynamic programming methods for assembly line balancing. Operations Research 30:375–390

Klingel A (1966) Bias in PERT project completion time calculations for a real network. Management Science 13:B194–B201

Kolisch R, Sprecher A, Drexl A (1995) Characterization and generation of a general class of resource-constrained project scheduling problems. Management Science 41:1693–1703

Kuchta D (2001) Use of fuzzy numbers in project risk (criticality) assessment. International Journal of Project Management 19:305–310

Lipke W (2003) Schedule is different. The Measurable News Summer:31–34

Lipke W (2004) Connecting earned value to the schedule. The Measurable News Winter:1, 6–16

Lipke W (2006) Applying earned schedule to critical path analysis and more. The Measurable News Fall:26–30

Lipke W, Zwikael O, Henderson K, Anbari F (2008) Prediction of project outcome: The application of statistical methods to earned value management and earned schedule performance indexes. International Journal of Project Management Forthcoming

Loch C, De Meyer A, Pich M (2006) Managing the unknown: A new approach to managing high uncertainty and risk in project. John Wiley and Sons, Inc., New Jersey.

Martin J (1965) Distribution of the time through a directed, acyclic network. Operations Research 13:46–66

Mastor A (1970) An experimental and comparative evaluation of production line balancing techniques. Management Science 16:728–746

Moder J, Phillips C, Davis E (1983) Project Management with CPM, PERT and precedence diagramming, chapter 10, 3rd edition. Van Nostrand Reinhold, New York

Pascoe T (1966) Allocation of resources - CPM. Revue Française de Recherche Opérationnelle 38:31–38

Patterson J (1976) Project scheduling: The effects of problem structure on heuristic scheduling. Naval Research Logistics 23:95–123

PMBOK (2004) A Guide to the Project Management Body of Knowledge, Third Edition. Newtown Square, Pa.: Project Management Institute, Inc.

Pollack-Johnson B, Liboratore M (2005) Project planning under uncertainty using scenario analysis. Project Management Journal 36:15–26

Roemer T, Ahmadi R (2004) Concurrent crashing and overlapping in product development. Operations Research 52:606–622

Roemer T, Ahmadi R, Wang R (2000) Time-cost trade-offs in overlapped product development. Operations Research 48:858–865

Schonberger R (1981) Why projects are "always" late: A rationale based on manual simulation of a PERT/CPM network. Interfaces 11:65–70

Schwindt C (1995) A new problem generator for different resource-constrained project scheduling problems with minimal and maximal time lags. WIOR-Report-449 Institut für Wirtschaftstheorie und Operations Research, University of Karlsruhe

Shtub A (1991) Scheduling of programs with repetitive projects. Project Management Journal 22:49–53

Shtub A, LeBlanc L, Cai Z (1996) Scheduling programs with repetitive projects: A comparison of a simulated annealing, a genetic and a pair-wise swap algorithm. European Journal of Operational Research 88:124–138

Steward D (1981) The design structure system: A method for managing the design of complex systems. IEEE Transactions on Engineering Management 28:71–74

Tavares L (1999) Advanced models for project management. Kluwer Academic Publishers, Dordrecht, 1999.

Tavares L, Ferreira J, Coelho J (1998) On the optimal management of project risk. European Journal of Operational Research 107:451–469

Tavares L, Ferreira J, Coelho J (1999) The risk of delay of a project in terms of the morphology of its network. European Journal of Operational Research 119:510–537

Tavares L, Ferreira J, Coelho J (2002) A comparative morphologic analysis of benchmark sets of project networks. International Journal of Project Management 20:475–485

Tavares L, Ferreira J, Coelho J (2004) A surrogate indicator of criticality for stochastic networks. International Transactions in Operational Research 11:193–202

Thesen A (1977) Measures of the restrictiveness of project networks. Networks 7:193–208

Van Slyke R (1963) Monte Carlo methods and the PERT problem. Operations Research 11:839–860

Vandevoorde S, Vanhoucke M (2006) A comparison of different project duration forecasting methods using earned value metrics. International Journal of Project Management 24:289–302

Vanhoucke M (2008a) The effect of project schedule adherence and rework on the duration forecast accuracy of earned value metrics. Tech. rep., Ghent University

Vanhoucke M (2008b) Improved project time performance using activity sensitivity and network topology information. Tech. rep., Ghent University

Vanhoucke M (2008c) Measuring time using novel earned value management metrics. In: Proceedings of the 22nd IPMA World Congress (Rome), vol 1, pp 99–103

Vanhoucke M (2008d) On the dynamic use of project performance and schedule risk information during project monitoring. Tech. rep., Ghent University

Vanhoucke M (2008e) Project tracking and control: can we measure the time? Projects and Profits August:35–40

Vanhoucke M (2009) Static and dynamic determinants of earned value based time forecast accuracy. In: Kidd T (ed) Handbook of Research on Technology Project Management, Planning, and Operations, Information Science Reference, forthcoming

Vanhoucke M, Demeulemeester E (2003) The application of project scheduling techniques in a real-life environment. Project Management Journal 34:30–42

Vanhoucke M, Vandevoorde S (2007a) Measuring the accuracy of earned value/earned schedule forecasting predictors. The Measurable News Winter:26–30

Vanhoucke M, Vandevoorde S (2007b) A simulation and evaluation of earned value metrics to forecast the project duration. Journal of the Operational Research Society 58:1361–1374

Vanhoucke M, Vandevoorde S (2008) Earned value forecast accuracy and activity criticality. The Measurable News Summer:13–16

Vanhoucke M, Vandevoorde S (2009) Forecasting a project's duration under various topological structures. The Measurable News Spring:26–30

Vanhoucke M, Demeulemeester E, Herroelen W (2001) An exact procedure for the resource-constrained weighted earliness-tardiness project scheduling problem. Annals of Operations Research 102:179–196

Vanhoucke M, Coelho J, Debels D, Maenhout B, Tavares L (2008) An evaluation of the adequacy of project network generators with systematically sampled networks. European Journal of Operational Research 187:511–524

Williams T (1992) Criticality in stochastic networks. Journal of the Operational Research Society 43:353–357

Williams T (1999) Towards realism in network simulation. Omega, The International Journal of Management Science 27:305–314

Yao MJ, Chu WM (2007) A new approximation algorithm for obtaining the probability distribution function for project completion time. Computers and Mathematics with Applications 54:282–295

Index

MIX
Papier aus verantwortungsvollen Quellen
Paper from responsible sources
FSC® C105338

If you have any concerns about our products,
you can contact us on
ProductSafety@springernature.com

In case Publisher is established outside the EU,
the EU authorized representative is:
Springer Nature Customer Service Center GmbH
Europaplatz 3, 69115 Heidelberg, Germany

Printed by Libri Plureos GmbH
in Hamburg, Germany